Changes in Innovation

Changes in Innovation

Towards an Improved Understanding of Economic Renewal

Edited by

Jani Saarinen

and

Nina Rilla

First published 2009 by
PALGRAVE MACMILLAN

Palgrave Macmillan in the UK is an imprint of Macmillan Publishers Limited,
registered in England, company number 785998, of Houndmills, Basingstoke,
Hampshire RG21 6XS.

Palgrave Macmillan in the US is a division of St Martin's Press LLC,
175 Fifth Avenue, New York, NY 10010.

Palgrave Macmillan is the global academic imprint of the above companies
and has companies and representatives throughout the world.

Palgrave®and Macmillan®are registered trademarks in the United States,
the United Kingdom, Europe and other countries.

ISBN 978–0–230–57744–2

This book is printed on paper suitable for recycling and made from fully
managed and sustained forest sources. Logging, pulping and manufacturing
processes are expected to conform to the environmental regulations of the
country of origin.

A catalogue record for this book is available from the British Library.

A catalog record for this book is available from the Library of Congress.

10 9 8 7 6 5 4 3 2 1
18 17 16 15 14 13 12 11 10 09

Printed and bound in Great Britain by
CPI Antony Rowe, Chippenham and Eastbourne

Contents

List of Figures

List of Tables

Notes on the Contributors

Juha Oksanen is a research scientist and the team leader of the VTT 'Innovation Policy Research and Impact Assessment of R&D' unit. His research interests include development of innovation policy, knowledge flows and user involvement in innovation processes, innovation in experience economy, and the links between entrepreneurship and innovation. He has been involved in the evaluation of national and EU programmes aimed to promote R&D and innovation, and public innovation service organizations.

Pekka Pesonen is a research scientist at VTT within the 'Innovation and Industrial Renewal' research team. His research interests pertain to various aspects of the innovations themselves, and to the link between innovation, technological development, and the competitiveness of firms and industries. His research has been mainly focused on innovation in and dynamics of the forest sector, and on innovation management issues, as well as managing SFINNO, the Finnish innovation-based database.

Nina Rilla works as a research scientist at VTT within the team 'Innovation and Industrial Renewal'. Her research interest lies in the field of industrial innovation, as well as on internationalization of companies and their innovation activities, and R&D in general. She has focused on investigating the changes in the innovation process, particularly from an innovation networks' and innovation failure's perspective.

Jani Saarinen works as innovation and technology policy manager at PricewaterhouseCoopers. Trained in both research policy and economic history, Saarinen studies innovation at the intersection of science, technology, and industry. His work includes extensive studies of industrial research and development, commercialization and business models of innovation, the development of manufacturing technology in Finland, and studies related to innovation and technology policies. In addition, he has expertise in innovation activities and technological evolution of firms and industries, long-term changes in the nature of R&D activities of firms and research centres, and also various issues related to patents and human capital.

Robert van der Have is employed as a research scientist at VTT within the team 'Innovation and Industrial Renewal'. His expertise is in economic-geography issues such as agglomeration externalities, industry clusters, and the measurement and analysis of spatial phenomena, especially innovative activity. His research interests include interorganizational knowledge creation in innovation, spatial evolution of industry and innovative activity, the interplay of regional environments and organizational dynamics, dynamics of entrepreneurship, as well innovation in services.

Preface

Thomas Alva Edison created several inventions in the field of telecommunications during the 1860s and 1870s. After realizing the commercial potential of his inventions, and with the help of the money he received in exchange for the rights for many of his patents, Edison decided to support and manage the creation of novel solutions more professionally. In 1876 he set up an R&D laboratory in Menlo Park, New Jersey, being the first of a kind in the USA, bringing together multidisciplinary knowledge and resources needed to facilitate innovation.

After a few years Edison introduced his most well-known achievement, the electric light bulb. Although Edison and his workers did not invent the first electric light bulb, with the help of a clear strategy of combining the existing knowledge and the capabilities of his laboratory, they created the first commercially viable incandescent light bulb. Edison's vision of electric lighting having the potential to be utilized by common people came to fruition after continuous improvement.

However, the lack of an energy supply and distribution system hindered the wider diffusion of innovation. Thus Edison needed to come up with new types of innovations supporting the core product. In 1878, he formed an electrical company in New York City, and couple of years later introduced a process innovation: the world's first economically viable system of centrally generating and distributing electric light, heat and power. This enabled the wide implementation of single light bulbs and can be regarded as the innovation having far greater impact on society than the light bulb used in the electrical system. Furthermore, it enabled Edison to create new services such as providing electricity to street lamps.

Despite the success in inventing and innovating, Edison encountered challenges and even failure, both in developing an innovation and in business. For instance, in the case of light bulb, Edison tested thousands of filaments before he discovered a carbonized bamboo filament that could last over 1200 hours. However, he saw failure quite differently, saying that, 'I have not failed. I've just found 10,000 ways that won't work.' In business, his idea to create a practical way to mine iron ore was never realized, leading Edison to lose the money he invested in the company behind the development. Yet, Edison can be regarded as a distinguished innovator. Not because of the many inventions he developed or of the many patents he possessed, bought and sold, but most of all because, unlike his peers developing new technology at the time, he focused on creating applicable and commercial solutions.

The electric light bulb is still widely used in household and commercial lighting, even though its product life cycle seems to be coming to an end. New product innovations, namely and mostly compact fluorescent lamps (energy-saving lights) and LED lights, are replacing the mature light bulb technology.

The case of Thomas Alva Edison elucidates well the complex nature of innovation and the various points of view from which it can be approached. Now, 130 years after Edison's first successful test of a light bulb with a carbon filament, the innovation phenomenon appears an ever more diverse and complex issue. However, the nature and many aspects of innovation illustrated in the case of Edison are still present and dealt with in this book; facets ranging from the influence of local circumstances, effort of learning and building up new knowledge, creating value from services, entrepreneurial endeavours and the like. Not surprisingly then, innovation as a phenomenon can be best understood from more than one perspective.

A multidisciplinary innovation studies tradition has been built at VTT Technical Research Centre of Finland during the past seventeen years. The research on which this book is based has been carried out over a four-year period in two wider research projects, which concentrated on examining changes in innovation and innovation processes. Not only have we been able to conduct interesting research on changes in innovation processes, but, significantly, we have been engaged with, and one could even say educated in, the field of innovation studies.

Without financial support from the Finnish Funding Agency for Technology and Innovation (Tekes) and VTT Technical Research Centre of Finland this book would have not been realized. We would specifically like to thank Eija Ahola and Anna-Maija Rautiainen from Tekes for their cooperation. In addition we owe sincere thanks to Sirkku Kivisaari from VTT and Raimo Lovio from Helsinki School of Economics for their valuable comments in steering our work over the years.

We also acknowledge the lively and fruitful discussions among our innovation studies colleagues at VTT and Helsinki University of

Technology that have provided us with valuable insights for research presented in this book. Moreover, the constructive comments of several commentators and reviewers in international conferences are highly appreciated.

Juha Oksanen
Pekka Pesonen
Nina Rilla
Jani Saarinen
Robert van der Have

Part I
Introduction

1
Changes in Innovation

Jani Saarinen and Nina Rilla

From the very beginning of mankind, technology has been one of the most essential and most important factors for the development of society. During the last 200 years, technological change has often been related to economic growth in the form of new types of goods and services. However, until the early days of the 20th century, technology was seen as something exogenous, which appeared outside the economic system and forced the economy to adjust itself to this new situation, in order to achieve an equilibrium. After some major advances made by such as Joseph Schumpeter in the area of technological development, innovations and industrial dynamics, technology was drawn inside the system and handled as something endogenous.

In the new growth theory, innovative activity has been considered as the engine of economic growth. Innovative activity is indirectly indicated by the Solow residual, that is, technological change. Technological change is the most important factor of economic growth in developed economies: new technologies explain at least half of the economic growth, maybe even three-quarters (Tassey, 1997). Technological change is driven by innovations. At the firm level, a constant flow of innovations is required to maintain competitiveness and organic growth.

One of the most recent trends in advanced economies has been an accelerating rate of innovations and technological change. Previously, innovation was often generated through a series of discrete steps in research, development and production. Today, innovation is increasingly generated in networks where value is generated through productive working relationships or collaboration (Tether, 2002). There has been a growing interest in innovation, entrepreneurship and technological change, and their impact on regional and national economic development and welfare. It is generally accepted that innovation is a major, if not the most important, source of productivity growth and that research and development (R&D) as innovation input is important. However, when it comes to knowing in precise detail the interrelations between R&D, innovation and productivity growth, we get less clear answers. For one thing, it is very difficult to distinguish innovations and R&D from other sources of productivity growth (Kenny and Williams, 2001). It seems that statements about the importance of innovation and R&D for productivity growth are based on calculations requiring strong simplifying assumptions or on arguments of a long-term historical and common-sense character. Another reason is the lack of an appropriate innovation measure.

Studies presented in this book belong to a wider research programme, which contributes to a more comprehensive account of the changes in innovation processes and characteristics of innovations, and their relations to productivity, economic development and welfare in general. This has been performed with the help of databases on the output of Finnish innovations since 1945 and comparative analyses of innovation and economic growth. The programme broadly has its theoretical base in Schumpeter's main ideas of economic development, where the appearance of innovations, which are unevenly distributed over time across firms and industries, causes economic growth through cyclical fluctuations (Schumpeter, 1911). This argument has been further developed in, for instance, Dahmén (1950), where the concept of development blocks plays a central role, and Schön (1998; 2000; 2006), with the theory of transformation, rationalization and crises. One common denominator of this line of evolutionary economic research is the attempt to link the micro-foundations of the economy to macro-level development. The present studies contribute to this by analyzing the characteristics of innovations as well as changes in innovation processes over time. The wide topic is approached from various research perspectives and methodologies.

However, a serious constraint on evolutionary economic type of research is the scarcity of empirical evidence considering the output and characteristics of innovations. In particular, there are few long-term studies. Much noticed is the British SAPPHO database, covering close to 4400 innovations during the period 1945–83, though considering only a few characteristics (Townsend *et al.*, 1981; Pavitt, 1984; Freeman and Soete, 1997; Tether *et al.*, 1997). Saarinen (2005) extended the Finnish innovation database (SFINNO), developed at VTT Technical Research Centre of Finland (Palmberg *et al.*, 2000) and originally starting with the year 1985, backwards in time to 1945. Consequently, the knowledge about innovations and their role in economic growth can greatly benefit from the innovation database such as SFINNO and the multidisciplinary research around it. Innovation-based data provides both qualitative and quantitative information on innovations, which can be cut down to the various categories related to innovations and their development processes. However, the largest advantage of the innovation database is the possibility to study long-term changes in the interrelations of various characteristics of innovations and the economic development of the country. As shown by Saarinen (2005), Finnish innovations have displayed a varying pattern, and not a unidirectional trend, as regards crucial characteristics such as development time and the age of innovating firms. This is an important finding and emphasizes the significance of a long-term approach used in the research programme.

Introduction to the chapters

There are five different parts in this book. In the first part we introduce the main objectives of the book and go through the Finnish innovation data SFINNO in more detail. The SFINNO data is used in several articles of this book, so we consider it important to present the data already here in the beginning of the book.

The second part is entitled as 'Characteristics of Innovation'. We start this part by taking a closer look at professionalization of research and development activities in Finnish companies before the 1970s. Chapter 3 leads us to the origins of Finnish industrial innovations, to those circumstances and situations, where new ideas and products have originated at that point of time. In Chapter 4 we study the existence and nature of innovation in mature industry. We address the specific question of whether mature industry is innovative and, if so, what the focus of innovation is in the particular stage of the industry life cycle. Chapter 5 concentrates on the question 'Are there failing innovations?'. This chapter searches for understanding of failure in innovation activities, aiming to analyze the understandings given for innovation failure.

In the third part of this book we concentrate on sectoral aspects of innovations. We start with Chapter 6 about innovation and dynamic strategy – planning and implementing continuous renewal. This chapter argues that while innovation is a central ingredient for firm success, in order to guarantee triumph in the longer run innovation needs to be understood in a wider organizational context. Chapter 7 is a study about spatial changes in innovation processes over time in Finland, in which we analyze long-term changes in the relationships between various characteristics of innovations and the spatial changes in innovation processes in Finland. This part is concluded by Chapter 8 in which we approach the topic of convergence in innovation from various points of view. The fourth part of this book analyzes the recent trends in innovation. We start with Chapter 9 about the international dimension of innovation process, using evidence from Finnish innovation data. Here the author explores the degree of collaboration in Finnish innovations in general, and the international dimension of the innovation process in particular. After analysis of the international dimension, we go back to analyzing the failure theme. In Chapter 10, entitled 'When the Going Gets Tough: Failure of Innovative Businesses', the authors explore how the characteristics of single innovations, their individual innovation processes and the firms that commercialize them can be related to strictly defined business failures. The concluding chapter in this part frames the elements of service innovation, trying to advance such a framework, incorporating tangible and non-tangible dimensions of innovation in services.

The last part of this book, entitled 'Epilogue', draws together the main findings and conclusions of the chapters and discusses the challenges and possibilities in innovation studies in general by presenting researchers' perspectives on this topic.

Towards an improved understanding of economic renewal

This book is a collection of chapters all contributing to a theme of changes in innovation and innovation processes, and aims to create new knowledge about crucial issues in the Finnish and global innovation environment which forms critical intelligence in innovation strategies for firms and decision-makers. We believe that this book and the individual chapters in it will increase the level of understanding of the current trends in innovation research, like the tension between regional and globalized knowledge systems, technology development (such as complexity and global business formation from innovations) as well as critical issues of failures in innovation. Thus, we feel that the book will contribute in three distinct ways to better understanding the economic renewal. First, it will help decision-makers and industry to apply a more pro-active approach in innovation and innovation policy. Second, it significantly strengthens the small and fragmented academic field of innovation studies, particularly in Finland but in other countries as well. Finally, the book creates awareness and improves knowledge of a wide range of aspects of innovation for business actors which are in support of developing their innovation activities.

References

Dahmén, E. (1950) *Svensk industriell företagarverksamhet – Kausalanalys av den industriella utvecklingen 1919–1939, Band I & II (Entrepreneurial Activity in Swedish Industry in the Period 1919–1939)*, Uppsala: Almqvist & Wiksells Boktryckeri AB.

Freeman, C. and Soete, L. (1997) *The Economics of Industrial Innovation*, 3rd edn, London and Washington: Pinter.

Kenny, C. and Williams, D. (2001) 'What Do We Know About Economic Growth? Or, Why Don't We Know Very Much?', *World Development*, Vol. 29, No. 1, 1–22.

Palmberg, C., Niininen, P., Toivanen, H. and Wahlberg, T. (2000) 'Industrial Innovation in Finland', VTT Group for Technology Studies, Working Papers No. 47/00, Espoo: VTT.

Pavitt, K. (1984) 'Sectoral Patterns of Technical Change: Towards a Taxonomy and a Theory', *Research Policy*, Vol. 13, No. 6, pp. 343–73.

Saarinen, J. (2005) 'Innovations and Industrial Performance in Finland 1945–98', *Lund Studies in Economic History*, Vol. 34, Stockholm: Almqvist & Wiksell International.

Schön, L. (1998) 'Industrial Crises in a Model of Long Cycles: Sweden in an International Perspective', in T. Myllyntaus (ed.), *Economic Crises and Restructuring in History*, Katharinen: Scripta Mercaturae Verlag.

Schön, L. (2000) *En modern svensk ekonomisk historia – Tillväxt och omvandling under två sekel (Modern Swedish Economic History – Growth and Change During Two Centuries)*, Stockholm: SNS Förlag.

Schön, L. (2006) *Tankar om cykler, Perspektiv på ekonomin, historien och framtiden (Thinking in Cycles: Perspectives on Economy, History and Future)*, Stockholm: SNS Förlag.

Schumpeter, J. (1911) *Theorie der wirtschaftlichen entwicklung*, Leipzig: Duncker & Humboldt. English translation, *The Theory of Economic Development*, Harvard, 1934; 8th edn, 1968, Cambridge, MA: Harvard University Press.

Tassey, Gregory (1997) *The Economics of R&D Policy*, London: Quorum.

Tether, Bruce (2002) 'Who Co-operates for Innovation, and Why: An Empirical Analysis', *Research Policy*, Vol. 31, No. 6, pp. 947–67.

Tether, B., Smith, I. and Thwaites, A. (1997) 'Smaller Enterprises and Innovation in the UK: the SPRU Innovations Database Revisited', *Research Policy*, Vol. 2, 19–32.

Townsend, J., Henwood, F., Thomas, G., Pavitt, K. and Wyatt, S. (1981) 'Science and Technology Indicators for the UK: Innovations in Britain since 1945', SPRU Occasional Paper No. 16.

2
Innovation as Objective: The SFINNO Approach

Robert van der Have, Jani Saarinen, Pekka Pesonen and Nina Rilla

Introduction

Research on innovations and innovative activity has been applying different sources of information to help in understanding the phenomenon and to investigate various research questions related to it. As the field of innovation studies is quite broad, incorporating various research topics which, in addition, can be approached from various perspectives, there cannot be one source of data superior to all others. Rather the different databases and information repositories offer diverse data, and thus are often taken as complementing rather than substituting each other. Most commonly used information sources include, for instance, patent databases and patent citation databases, innovation and R&D surveys, as well as R&D expenditure statistics. In the chapters of this book, the main source of innovation data has been the database of Finnish innovations, namely SFINNO. The database is developed, constructed and maintained by VTT Technical Research Centre of Finland.

The SFINNO database includes a diverse set of data constructed on the basis of single innovations. The data is delineated roughly on three levels: the innovation (the commercialized output of the development process); the innovation process (the actual process in which the original idea developed to a commercialized innovation, and further); and the innovating firm (the organization mainly responsible for the development and commercialization of the innovation). As the base unit of the data is innovation, the information on the above-mentioned three levels is assorted in relation to a single, identified innovation. Today, the database includes 4537 innovations in total, which have been commercialized between 1945 and 2007.

In this chapter, the SFINNO database is described in more detail in order to give a more profound understanding of the data used in the following chapters so as to help the reader in interpreting the results. In addition, we, the authors, believe that as the database has many advantages in innovation studies, it is needful to disseminate these benefits to the awareness of other researchers to help in creating and developing innovation data sources, and consequently push further the forefront of the research field. We begin by discussing the primary ways used to collect innovation data, and then proceed to describe the construction and contents of SFINNO, followed by a description of the present state of the database.

Two approaches in gleaning data on innovations

As collecting innovation data can be approached via two principally different ways, we will briefly describe these, so as to make clear the value added of the far less common methodology used for the SFINNO database. At the outset, larger-scale collection of (non-technometric) quantitative innovation data may be approached via two distinct perspectives: 1) that of the innovator (commonly referred to as the *subject approach*) and 2) that of the single object of innovation (i.e., the innovation itself, hence labelled the *object approach*).

The subject approach has been the most widely applied and most influential method in studies of innovation, usually taking the form of an innovation survey (most notably the European Community's and USA's Innovation Surveys). Despite its wide acceptance and application, the subject approach has displayed some characteristics that need to be critically considered. With the subject approach, one does not collect explicit data on any specific innovation. Instead, one collects general (typically aggregated) data on the firm level regarding its innovations and innovative activities (inputs, outputs and characteristics such as collaboration). It has been common practice in such surveys to measure innovative *activity* in the firms through proxies such as R&D or innovation expenditures (as a type of input) and patents (as a type of output). Therefore, the subject approach is able to address more general innovation questions at the firm, sector and country level, but not on the level of specified and isolated innovation processes (i.e., linked to single innovations) or individual innovations. As such, information on the nature of innovations that were introduced by the subjects is not well captured by the subject approach.

In contrast, the object approach takes a specific innovation as point of departure by counting and recording individual innovations, and thus collects the data that can be related to these innovations. Dealing directly with the output of an innovation process, it can provide more detailed information on the characteristics of the innovation itself, the underlying knowledge base, its origin and development (Palmberg *et al.*, 1999). Here, some similarities are shared with patent analysis, but a main difference is that, although counting pure innovations only, the object approach cannot cover the whole population of innovations introduced to the economic system (Archibugi and Pianta, 1996). Hence, due to differences in the likelihood of being picked up in the research environment between heterogeneous innovations, a degree of bias towards innovations which are perceived to be more relevant exists. This inhibits the use of representative samples for the entire population of innovations. Nevertheless, rich data can be collected on a wide range of innovations. One more specific advantage of the object approach is that any following information may be still retrieved from the subject (the innovating firm), but with the additional strength that this enquiry is specifically targeted, which helps to get more focused and therefore more reliable additional information.

Within the object approach, two different (but non-exclusive) methods have been used to identify the innovations. The first method is the use of expert opinions for identification of innovations, which is followed by tracking of the commercializing firm and either directing a survey to acquire additional data or the use of publicly available data sources. The second method which has been applied identifies the innovation via systematic reviewing of technical and trade journals, and records the basic data on the innovations from these, without the use of an additional survey directed to the commercializing firm. This method is known as the literature-based methodology or literature-based innovation output (LBIO) method. The main justification for using the LBIO method is the fact that editors of professional trade and technical journals have the economic incentive to publish about relevant or commercially 'interesting' innovations. This follows the same logic as the use of patent data, as inventors have an economic incentive to patent their invention for the obtainment of temporary monopoly rights. Hence, journals may be functional as a relevant source of innovation data (Palmberg *et al.*, 1999).

One significant and early application of the object approach was carried out by the Science Policy Research Unit (SPRU) of the University of Sussex, UK. For the construction of the SPRU innovation database, the expert opinion method was used, involving nearly 400 experts from a wide range of research and trade organizations, government, the academic world and trade and technical journals, firms and consultancies

(Palmberg *et al.*, 1999). This effort has lead to a number of successful empirical studies based on the database (see, e.g., Pavitt, 1984). One other example of the use of expert opinion method is a database constructed by the Gellman Research Association in the USA in the 1970s which was initiated by the National Science Foundation for the purpose of developing science and technology indicators (see Acs and Audretsch, 1993, for a brief description).

In comparison with often-used data sources, like patents and CISs (Community Innovation Surveys), the object approach of identifying and analyzing single innovations shows various advantages. First, patents are not equal to innovations as not all patents can be considered innovations (i.e., commercialized technology) and, on the contrary, not all innovations are patented (see Mäkinen, 2007). Furthermore, the linkage between a patent and an innovation is challenging as one innovation can resemble several patents.[1] Thus patents are not a straightforward measure of innovations. In addition, only technological novelties can be patented, yet innovations can take a non-technological form as well. It is increasingly important to study innovations in services and in organizations, and patent data is unable to meet this need. For the purpose of analyzing non-technical renewal one can apply trademarks or utility patents, but the problem remains that they are not necessarily comparable to single innovations. Second, to analyze companies' innovation activity, development of various types of innovations and countries' innovativeness, CIS is applied. However, this subject approach provides data on the firm level and from the point of view of the firm (innovator) as it is based on a survey sent to a sample of companies. Thus, for instance, the definition of innovation ultimately depends on the opinion of the survey respondent resulting in data with heterogeneous definitions, making the actual focus of measurement ambiguous. Object approach has the benefit of the possibility of having preset definitions for the identification of innovations, yet including innovations from a variety of firms and industries. For instance, the object approach enables the inclusion of micro firms (fewer than ten employees), which has been noted to be highly relevant among innovative firms (see Saarinen, 2005), but which, for example, CIS excludes.

The method and contents of SFINNO

SFINNO basic data

The first activity for SFINNO was to produce a database consisting of 1673 Finnish innovations commercialized in the 1980s and 1990s. The whole

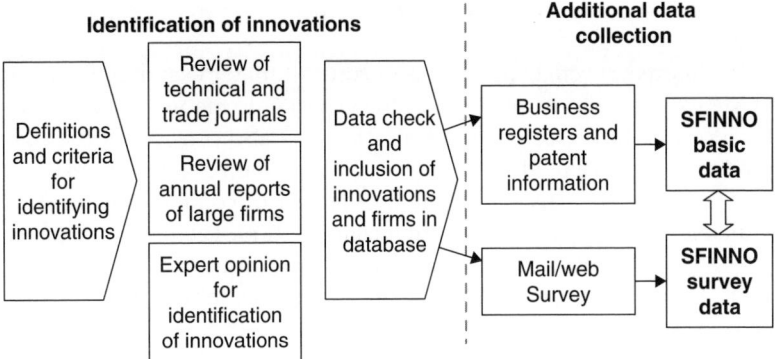

Figure 2.1 SFINNO data collection process

dataset was compiled using a combination of three different methodologies for the identification of innovations (Figure 2.1): 1) expert opinion; 2) reviews of trade and technical journals; and 3) reviews of the annual reports of large firms. Of these three, expert opinion and literature-based reviews are relatively well-established methodologies in innovation studies similar to SFINNO. The reviews of the annual reports of large firms, however, take a somewhat different point of departure since the innovations were identified through a subjective judgement in collaboration with the firms. The use of expert opinion for the identification of innovations at VTT began in 1992. This involved more than 150 experts representing different industrial and technological fields from industry, VTT, Tekes[2] and the universities of technology in Finland. These experts were asked to list significant innovations according to certain definitions and criteria and to identify the year of commercialization and the commercializing firm. This exercise resulted in the identification of an additional 285 innovations.

To conduct the literature reviews, first, the population of journals that were eligible for innovation detection was defined. Journals were considered eligible if they were independently edited and regularly published; that is, mere product listings or announcements, irregular publications or journals directly controlled by companies were not considered eligible. Subsequently, those journals were selected that regularly published edited and non-paid material about innovations. The focus was on articles dealing with the introduction of new products which conformed to the definitions and criteria for an innovation. Listings of new products were avoided. Instead, more emphasis was paid on the editorial content of the journals. Altogether 18 journals fulfilled these criteria.

This more restricted approach is in contrast to, for example, the work of Kleinknecht and Bain (1993) or the OECD Oslo manual's[3] guidelines for literature-based innovation collection that also consider non-edited product announcements to be eligible. Identified innovations were listed and described by SFINNO, including the year of commercialization (if available), the name of the commercializing firm, the journal number and the relevant pages. This resulted in the identification of 1144 innovations. Furthermore, lists of award-winning innovations in the literature reviews were also included.

To pay special attention to large firms, as is done in SFINNO, is also in contrast to previous LBIO studies. In the studies conducted by the Futures Group in the USA (based on the SBIDB data)[4] and by Acs and Audretsch (1990), a common concern was the question whether large firms' innovations were representative. A characteristic of these studies was that they collected data in the so-called 'new product' sections in magazines, whereas in SFINNO these sections are avoided.

A common problem in object approach studies is that the identification of innovations has not been based on statistical sampling, since the population of all innovations remains unknown. In subject-based studies, like CISs, the statistical significance of the sample has been emphasized. In order to get some hints about the validity and reliability of the SFINNO basic data one can compare the collected data with other data. To this end, the contents of the SFINNO data have been compared with Finnish patent data[5] and with the Finnish CIS data. By and large, innovation outputs as identified by the SFINNO approach compared with patent counts had a similar sectoral distribution (the software sector being excluded due to patentability issues). According to one comparison, in general, the object-based SFINNO and subject-based CIS datasets produce consistent results as far as major characteristics of innovators and innovation processes were compared.[6] It is worth noting that similar results were found in Acs *et al.* (2002).

Due to the importance of a few large firms in the Finnish economy, these firms were included on a case-by-case basis (initially altogether 24 firms), since it was feared that they would otherwise not be sufficiently covered and in enough detail through the literature reviews. The selection of firms was also made on the basis of R&D spending and patenting, as it was assumed that firms investing heavily in R&D and patenting could also be considered innovative. Hence, in the construction of SFINNO, all new products that these firms had launched between

1985 and 1998 were listed. The firms were then approached with the lists of product launchings and the definitions and criteria of an innovation, and they were asked to pick out those products which they considered especially important and innovative. In this way a group of 200 innovations were identified. Another group of 226 innovations commercialized between 1985 and 1998 has been identified more or less unsystematically from miscellaneous written sources, the internet or by researchers.

The combination of different methodologies for the identification of innovations was intended to secure a good coverage of the data across different industries and firm sizes. On the other hand, this also implies that it is more difficult to control for bias of any kind. Biases may arise, for example, if the experts were inclined to identify relatively more innovations originating from bigger firms, or if the literature reviews identified relatively more innovations from smaller firms. Another bias in favour of innovations from larger firms may arise through the review of annual reports on a case-by-case basis.

In order to check for bias, a cross-comparison between the sources of the identification of innovations was carried out. The result suggests that relatively more innovations from smaller firms were identified through literature reviews. On the other hand, the experts did not have a significant bias in favour of innovations from bigger firms. Moreover, the share of innovations which were identified from more than one source is relatively small, indicating that the combination of different methodologies has indeed enhanced the representativeness of the database. Another problem related to the identification of innovations from different sources is double-counting. In order to avoid this, each innovation was checked for duplicates before they were added to the database.

The information about innovations in the SFINNO basic data was completed through the identification of the innovating firms and the collection of information about these firms from two main sources: company registers and patent databases.[7] The world wide web was also used as a complementary source. Apart from identifying and including innovations in the database, additional sources were used in order to ensure at least certain basic data on innovations.

SFINNO survey data

In addition to the data gathered from public sources, a survey is sent to firms in order to get more specific information of the individual innovation processes. The first mail survey considering innovations from 1985–98 was undertaken in four successive phases between December 1998 and October 1999. The response rate for mail survey reached

67 per cent. Of the 1235 questionnaires posted, 729 were returned. The survey data in practice covers only active firms, even though the innovations might already have exited the market.

The SFINNO survey questionnaire covers a wide range of aspects relating to the different phases in the process, from idea to commercialized innovation and further. These topics include origin and drivers of the innovation, funding, collaboration, patenting, exporting and internationalization of the innovation, innovation diffusion, commercial success, timeliness of the process, novelty of the developed innovation, as well as challenges in and impacts of the innovation for the commercializing firm.

Besides the innovation data compiled for innovations commercialized between 1984 and 1998, actions to get data from further back in history were taken. During 2001–2, a literature review for years 1945–84 was conducted. The result of this exercise was the creation of complete time series data for characteristics of innovations and innovation processes for the years 1945–98. The total number of innovations collected for this particular period was some 3100, and the number of developing and commercializing Finnish firms reached 1700. The collected innovation-related variables include information origin of innovation, role of customers, collaboration, public funding and so on. For each innovation, there is also specific information on the commercializing firm. Again, an innovative firm has been defined 'as a firm, which has developed and commercialized a new product – an innovation'. In order to get the historical data comparable with the already existing data, the same variables, with some small modifications, were collected and included in the database. Hence, the main difference between the innovation data for the periods of 1945–84 and 1985–98 is the collecting process itself. For 1985–98, the so-called holistic approach was used, which means that several different methods were used for identifying innovations. With regard to the historical (1945–84) data, only the literature-based method was applied.

The same criteria which were used in the collection process of the SFINNO basic data were applied for the historical part as well. After the selection process of relevant journals, the final list included 36 journals out of 42. The large number of 36 is caused by the changes in the name of the journal over time. As it turned out, the list of journals for the historical period is quite similar compared to the list of journals in the original SFINNO. Together, the selected journals can be considered a good sectoral coverage of industrial life in Finland during this particular period.

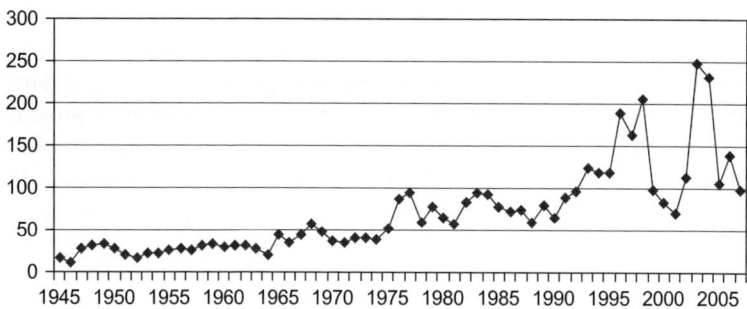

Figure 2.2   Count of innovations in SFINNO according to the commercialization year

In addition, annual reports were also reviewed for the historical part. However, this turned out to be somewhat problematic for the identification of innovations, and therefore it was decided to complement the annual reports with company histories to get higher information content. The number of innovations between 1945 and 1984 then reached 2208. After a check for duplicates, the number of innovations in the historical part was reduced to 1593. The different phases of the collection process are well documented (see Saarinen, 2005). For the historical part, data on commercializing firms was gathered from various sources, including the Finnish Patent and Trademark Office (PRH) and company histories. Enterprise data included information like year of establishment, Standard Industrial Classification code, location and financial information (turnover, balance sheet, employee costs).

Recent developments and present state

At present, the SFINNO covers data on innovations commercialized up to the year 2007 (see Figure 2.2) and follows the same robust method. The sources for the identification of innovations, as well as the definition of innovation, have been kept the same, even though expert opinions have not been collected systematically. Today, 15 journals are used in identifying innovations and annual reviews of 15 large firms are used in completing the annual set of innovations. The most recent survey round was executed during the summer of 2008 and the according data was added to the database in autumn 2008.

In 2007–8 a number of actions were taken to develop and refine the database towards a better and more reliable applicability. These actions included updating of company exit data, reviewing the location of innovation development of large (multi-location) companies, deleting a set of duplicate entries and linking longitudinal and more profound firm data (including firm size data over time and data on establishing process). The survey questionnaire was also elaborated further to better represent the topical issues in innovation studies at the moment, yet keeping the comparativeness to the existing survey data. Furthermore, a new extension of the SFINNO basic data and subsequent survey is in progress to cover service innovations as well. This follows efforts in conceptual work, which turned out to be necessary due to the pervasive and intangible nature of service products. Furthermore, efforts are being made by the VTT Innovation Studies to start databases based on the same instruments abroad, which would allow for comparative research in this vein.

Notes

1. Although patent families can be regarded as being closer to single inventions, and thus innovations, in some cases.
2. The Finnish Funding Agency for Technology and Innovation.
3. OECD Oslo manual (1997).
4. See discussion in Edwards and Gordon, 1984, pp. 14–15.
5. Palmberg *et al.*, 2000.
6. Leppälahti, 2000.
7. These data sources were available in an electronic form. The basic data on companies were provided by Statistics Finland. In case of information on turnover and number of employees, the data provided by Statistics Finland was combined with the data from the 'Voitto' CD-ROM, published by Asiakastieto. Considering the patent data, the Patent and Trademark Office in Finland was the origin of information.

References

Acs, Z. and Audretsch D. (1990) *Innovation and Small Firms*, Cambridge, MA: MIT Press.

Acs, Z. and Audretsch, D. (1993) 'Analysing Innovation Output Indicators: The US Experience', in A. Kleinknecht and D. Bain (eds), *New Concepts in Innovation Output Measurement*, London: Macmillan.

Acs, Z., Anselin, L. and Varga, A. (2002) 'Patents and Innovation Counts as Measures of Regional Production of New Knowledge', *Research Policy*, Vol. 31, 1069–85,

Archibugi, D. and Pianta, M. (1996) 'Measuring Technological Change Through Patents and Innovation Surveys', *Technovation*, Vol. 16, No. 9, 451–68.

Edwards, K. L. and Gordon, T. J. (1984) *Characterisation of Innovation Introduced to the US Market*, Report to the US Small Business Administration, Glastonbury: Futures Group.

Kleinknecht, A. and Bain, D. (eds) (1993) *New Concepts in Innovation Output Measurement*, London: Macmillan.

Leppälahti, A. (2000) 'Comparison of the Finnish Innovation Surveys', *Science Technology and Research 2000*, 1, Helsinki: Statistics Finland.

Mäkinen, I. (2007) 'To Patent or Not to Patent? An Innovation-level Investigation of the Propensity to Patent', VTT Publications 646, Espoo: VTT.

OECD (1997) *Proposed Guidelines for Collecting and Interpreting Technological Innovation Data: 'The Oslo Manual'*, Paris: OECD and Eurostat.

Palmberg, C., Leppälahti, A., Lemola, T. and Toivanen, H. (1999) 'Towards a Better Understanding of Innovation and Industrial Renewal in Finland: A New Perspective', VTT – Technical Research Centre of Finland, Working paper 41/99, Espoo: VTT.

Palmberg, C., Niininen, P., Toivanen, H. and Wahlberg, T. (2000), Industrial Innovation in Finland', VTT Group for Technology Studies, Working Papers No. 47/00, Espoo: VTT.

Pavitt, K. (1984) 'Sectoral Patterns of Technical Change: Towards a Taxonomy and a Theory', *Research Policy*, Vol. 13, 343–73.

Saarinen, J. (2005) 'Innovations and Industrial Performance in Finland 1945–98', *Lund Studies in Economic History*, Vol. 34, Stockholm: Almqvist & Wiksell International.

Part II
Characteristics of Innovation

3
Professionalization of Research and Development Activities in Finnish Companies before the 1970s

Jani Saarinen

Introduction

From a European point of view, industrialization got off to a relatively late start in Finland. Industrial production began in the 1860s and continued to grow at a steady pace up to the 1950s (Hjerppe and Vartia, 1998). During the post-war depression, industry suffered from a constant lack of raw materials and trained workforce. Insufficient funds made it impossible to invest in machinery and equipment to the extent required – not to mention the strict customs regulations that may have thwarted such purchases from abroad in the first place. Shortage of funding also affected industrial R&D, which was decidedly meagre in Finland compared to that of other industrialized countries. In 1956 research appropriations accounted for 0.2 per cent of the Finnish GDP, while the corresponding figure in more advanced countries was well over 1 per cent and over 2 per cent in the USA and UK (Törnudd, 1958). By the mid-1960s, R&D expenses had doubled to 0.4 per cent but were still at a low level in international terms (Finnish Government, 1974).

Despite these shortages, scientific and technical research was relatively extensive before the 1970s. The purpose of this chapter is to describe research in different industries at a time when Finland had not yet defined a goal-oriented policy for scientific and technical research, that is, a policy aiming to systematically promote science and technology (Lemola, 2001). In this context, scientific and technical research refers to activities that aim to 'enhance the security of national supplies and strengthen the country's economy to confront international competition' (Levón, 1958, 265–7). The main focus of attention is the period from 1945 to 1965, a time in which several industries saw the establishment of R&D 'organizations'. Most of these organizations were associations formed by different players in the field. Their main goal was to jointly develop technologies in the field, relying on relatively small financial resources and not aiming to generate profit. This chapter will discuss the conditions that had a broader impact on the establishment of these organizations in the post-war years. By combining the R&D data with the data of some 1000 Finnish innovations from the period before 1970, this study leads us to the origins of Finnish industrial innovations, to those circumstances and situations, where new ideas and products have originated.

To achieve these goals, the chapter will examine the following questions: 1) When and how did R&D activities begin in different industries? 2) Where was research carried out and how was it funded? 3) What kind of research was conducted and what were the results? The investigation focuses on 'traditional' industries, which played an important part in the post-war development of Finland. These include agriculture, forestry, the wood processing industry, construction industry and textile industry. This study does not deal with the scientific and technical research activities carried out at VTT Technical Research Centre of Finland[1] and KCL.[2]

Theoretical background

No ready theoretical framework exists for examining R&D activities carried out at the level of individual industries. To discuss the growth in R&D in Finland before 1970s, I have used two different approaches: one to examine the period, the other to study the strategic choices made by different players. In general, the growing need for R&D can be explained through the continuous accumulation of technological knowledge. This leads to an increase in the number of technical alternatives as a result of the multiplication of information. As Johnson and Striner (1960) point out, in the USA, for example, this meant the number of viable technical options increasing thousandfold in 1940–60. It is fair to assume that this information reached Finland a few years later. Second, the post-war years were the period of Ford's mass production (Freeman, 1987), characterized by an increasing use of automation and new technologies, as well as the standardization of components and materials. Along with the accumulation of technical competence, the emergence of new technologies after the war years provides at least a partial explanation for industry-specific R&D seeing such heavy growth in this period.

Porter's theory (1980) of the competitive strategies in emerging industries gives us added insight into the phenomenon. Although the industries discussed in this chapter were not emerging fields as such, they were all characterized by rapidly increasing automation. Industrial R&D was usually propelled by the need to solve a particular problem, usually a technical one. These problems did not trouble one company alone, but were often seen as a handicap in the entire field. The novelty of technologies and techniques allows us to call these fields 'emerging industries'. According to Porter, emerging industries are characterized, among other things, by their resulting from technological innovations. Emerging fields often experience a great deal of insecurity concerning the development of technology. Another characteristic in the early days is the large number of recently established companies. The more companies are involved, the less each company needs to contribute to the high start-up expenses. Cooperation enables companies and players to participate in technological development in the field with a relatively small initial investment.

National level

Research and development (R&D) is perhaps one of the most classical issues in studies concerning technological change and innovation activity of companies. Since the early 1960s, there has been no doubt in the economic literature about the importance of R&D activities in relation to the economic growth (Acs and Audretsch, 1988; Kleinknecht, 1996; Freeman and Soete, 1997; Griliches, 1995). One widely used measure of national investment in change-generating activities is the share of GDP spent on R&D activities. R&D expenditures are often seen as an input indicator in relation to the innovative process, whereas patents – and later on innovations – are regarded as an output indicator (OECD, 1986; Kodama, 1986). The R&D expenditure reflects formal expenditure on research and development as reported to the Central Bureau of Statistics. Therefore, the main advantage of this indicator is that data are available on firm, industry and national levels (OECD, 1993).

According to Freeman and Soete (1997), the distinctive feature in modern industrial R&D is its scale, its scientific content and the extent of its professional specialization. The professionalization in its turn is associated with three main changes: 1) the increasingly scientific character of technology; 2) the growing complexity of technology; and 3) the general trend towards division of labour, as noted already by Adam Smith. Usually, the developments in the USA after World War II have been mentioned as 'golden years' in terms of the emergence of company R&D, as particularly the role of small and new firms in the commercialization of new technologies – computers, semiconductors and biotechnology – increased (Bruland and Mowery, 2004). However, the early developments

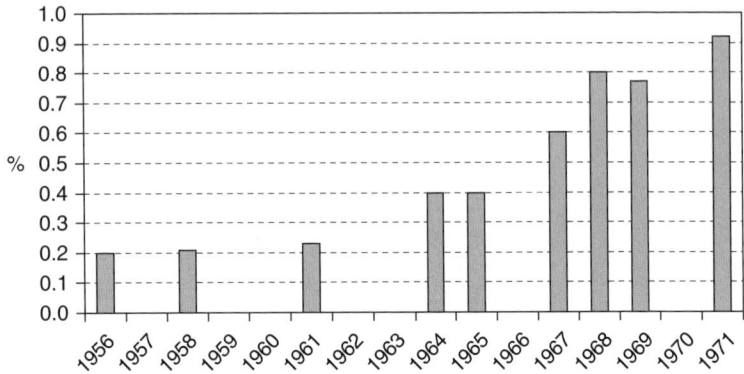

Figure 3.1   Distribution of private and public research and development expenditures, 1961–97 (in FIM billion)
Sources: State Council, 1974 [1962–72]; Elfvengren, 1961

in Finnish R&D laboratories are largely unknown in the literature. For an understanding of the development of innovations, I have seen it as necessary, since R&D significantly contributes to the innovative context, to reconstruct the early developments in the professionalization of Finnish R&D activities.

The first official statistics concerning scientific-technical activities were published in 1930s in the Soviet Union. A couple of decades later, in the 1950s, the Organisation for European Economic Co-operation (OEEC) made an attempt to construct systematic tools to measure research activities. Since 1963, the objective has been to use internationally standardized statistical tools in order to measure R&D expenditures. In Finland, a systematic collection of R&D statistics was started by Statistics Finland in 1969. However, already before that, some attempts were made to measure the research expenditures in Finland (Törnudd, 1958). Figure 3.1 displays the long-term development of private and public R&D expenditures in Finland.

Since the beginning of the period of research policy in the mid-1960s, R&D statistics have been one of the most observed indicators by Finnish policy-makers. New goals for the level of R&D expenditure have continuously been updated. As can be noticed from Figure 3.1, R&D expenditures have steadily increased. Despite a doubling of R&D in GDP from 0.2 per cent in 1956 to 0.4 per cent in 1965, R&D investments were relatively small in Finland until the mid-1960s (Törnudd, 1958). In the top three OECD countries (Great Britain, West Germany and the USA), the corresponding numbers were between 1.3 and 1.7 per cent of GDP already in the late 1950s (Törnudd, 1958). However, these low figures in Finland do not tell us the whole story about the level and richness of technical-scientific research there. A major role was played by different organizations and associations, which were established in order to solve certain specific problems in different industrial branches. The period before the mid-1960s witnessed a rapid increase of these problem-solving organizations (Saarinen, 2005), as a result of mechanization as well as rapid increase in the available technical and scientific knowledge. In the following paragraphs, the pioneering laboratories and problem-solving organizations are presented.

The very first laboratories in Finland were not established in order to develop new products for the markets, but for the solution of practical problems. A particular problem around 1900, when foreign trade increased continuously, was the rising variety of new items, which were imported to the Finnish market. The Customs Office had the responsibility to control the imported items, making Finnish consumers safe. In the late 19th century, the controlling and testing activities were done by the Laboratory for Agriculture and Trade Chemistry, but as the number of imported items kept increasing, the Customs Office's need for its own laboratory became actual. The Customs Laboratory was founded in 1908. Its main tasks since the beginning dealt with the chemical composition of imported goods.

Practical issues were also the background of the foundation, in the 1930s and 1940s, of a large number of forest-related research centres. A special feature of these laboratories was that they were established by a group of independent companies. The main reason for the cooperation was to share costs spent on research. These new laboratories had their own facilities, research personnel were mainly recruited from the universities, and they were financed by the founders. A common denominator for these new centres was that the research carried out by them was mainly concentrated on practical items, such as problems related to exports, weathering, strength qualities etc. When the Technical Research Centre of Finland (VTT) started its activities in 1942, some of these centres were transformed to VTT's wood-technical laboratories. However, only the mechanical research was covered by VTT, which meant that other types of forest- and wood-based research continued to be carried out in research centres like Keskuslaboratorio Oy (nowadays KCL Oy), Valtion Metsäntutkimuslaitos (Metla Oy), as well as Metsäteho (Metsäteho Oy).[3] Table 3.1 presents an overview about the research centres and laboratories established before the 1960s.

28

Table 3.1 Finnish research centres and laboratories (established before 1965)

Research centre (and sector)	Founded (year)	Established by (industry/ state/other)	Nature of R&D (basic/ applied)
Puutekniikan tutkimuksen kannatusyhdistys (paper- & wood-industry)	1929 Exit: -42	industry & state (later)	basic
Metsäteknologian tutkimusosasto (paper- and wood-industry)	1931	industry	basic
Työtehoseuran metsäosasto (paper- and wood-industry)	1942	private + later state	standardization
Suomen Puunjalostus-teollisuuden Keskusliiton metsätyöntutkimustoimisto (paper- & wood-industry)	1945	industry	applied
Metsäteho (1996: –> Ltd) (paper- and wood-industry)	1945	industry	applied
Uittoteho (paper- and wood-industry)	after the wars	industry	applied
Metsähallituksen hankintatoimisto (paper- and wood-industry)	after the wars	state	?
Suomen Villateollisuuden Tutkimuslaboratorio (textiles industry)	1947	industry	applied
Tiiliteollisuuden keskuslaboratorio (savi-, lasi- & kiviteoll.)	1948	industry	applied
Lumiauratoimikunta (vehicle industry)	1943	industry + state	basic & applied
Tekstiilitutkimus-yhdistys ry (textiles industry)	1954	industry	Applied int. research
Maatalouskoneiden tutkimuslaitos (agriculture)	1946	industry + state	applied R&D
Maatalouskoneiden tutkimussäätiö (agriculture)	1953	industry	applied
Pakkaustutkimus-laboratorio (Paris) (packaging industry)	early 1930s	industry	applied (?)
Pakkaustutkimus-laboratorio (K-lab.) (packaging industry)	1957	industry	applied
SAFA:n standardisoimislaitos (construction industry)	1950s	industry	standardization
Rastorin rakennusosasto (construction industry)	1950s	industry	applied
RAKEVA-säätiö (cosntruction industry)	1959	state + industry	coordination
Työtehoseuran maatalouskoulu & - tutkimuskeskus (agriculture)	1947	industry + money from state	basic + applied

According to Table 3.1, a large number of laboratories existed in the pulp and paper industry. This is quite understandable given how big a role the forest industry played in the Finnish economy during the studied period. In most of the cases, it was industry's initiative to start up laboratory activities. What usually happened was that a group of firms, which faced similar difficulties in some specific part of their production chain, were connected together to establish a joint venture for solving their problems. Researchers for these new 'firms' came from the universities and technical colleges. The type of research was mainly applied research, which focused on solving the daily problems of 'mother' companies. In the following section, a more detailed presentation about the activities of these laboratories is provided.

Sectoral level

Forestry and wood processing

Throughout history, forest resources have been a significant part of the Finnish national economy. Wide-ranging production and export in the sawmill and paper industries have been vital to the country's economic history. This makes it hardly surprising that R&D activities in the field got well under way in the years between the wars.[4] The biggest problems were related to the forests beyond the 'zero limit'. Owing to their location and the price and transport conditions at the time, these forests did not lend themselves to financially viable use (Pipping, 1955). Another setback came from the land cessions in 1944, which reduced Finland's forest areas and tree stock by some 15 per cent (Lindberg, 1956). Despite these problems, the mechanical wood-processing industry in Finland experienced a rapid and comprehensive technical revamp in the decade following World War II (Stenholm, 1956). Improved demand and transport conditions boosted forest use and improved their management in the late 1950s.[5] A new phenomenon in the field was the expanding cooperation in the Nordic countries, especially in the field of pulp and paper research (Serlachius, 1959). The wider availability of forest machinery, such as tractors and different types of saws, along with enhanced production methods, also furthered development (Vesterinen, 1952).

Wood technology research

The beginnings of Finnish wood technology research can be traced back to the need to solve blue staining, a problem that affected timber exports at certain times of the year. A support association was set up in 1929 to collect funds for research activities, and FIM 150,000 (nearly 25,230 euros) was allocated for the promotion of wood technology research in the 1930 national budget. Private funding was acquired in the form of support fees and donations from business and private members. Once under way, blue stain research focused on loading, sea transport and unloading, as well as the conditions that timber was subjected to during these phases. Later on, research teams began to test chemicals developed in the USA, Germany and Sweden to prevent blue staining (see, e.g., Roschier, 1959). Research showed that blue staining and the ensuing losses of up to FIM 50 million could be fully avoided with investments of FIM 2 million. In fact, once the investments had been made, blue timber pretty much disappeared from the market and sales catalogues (Jussila, 1952).

Other wood technology research projects included the field of artificial drying, supervised by Mr Sahlman (MSc). Sahlman eventually published a book on the results of several years of research and testing in the field. Another important research field involved the physical and strength properties of Finnish pine tree. Following thorough investigation, Professor Siimes published the findings in a doctoral thesis, which introduced several valuable, previously unknown principles governing the properties of timber. This work was followed by the establishment of strength grading as an approved standard. Research in the field was later transferred to VTT Technical Research Centre of Finland. The support association for wood technology research was merged with VTT after the Research Centre was established in 1942 (Jussila, 1952). Despite the important part played by the wood industry in post-war Finland, overall funds allocated to research in 1961 accounted for only 0.44 per cent of the added value of wood and paper industry products (Elfvengren, 1963).

Forest technology research

Finnish research on forest technology first focused on topics related to the use of small trees and studies concerning forest work. The establishment of a forest technology research unit in the Finnish Forest Research Institute in 1931 gave a boost to research on forest work. The primary task of the new unit was basic research, including, for example, studies on the structure and properties of wood and timber. Timber measurement studies were also part of the Forest Research Institute's tasks since the results were of public use and thus required research to be carried out by an impartial government research body. Despite the importance that this type of research carried to the entire forest industry, the shortage of researchers and funds kept research at a modest level.[6]

After the post-war lull, several private research bodies were set up in parallel with the Finnish Forest Research Institute for short-term, practical forest technology research projects. These included the department of forestry at the TTS Work Efficiency Institute, the department for forest work research at the Finnish Forest Industries Federation, Metsäteho, Uittoteho and the procurement department of Metsähallitus. The projects carried out by Metsäteho focused on, for example, the applicability of forest machinery, development of new methods for logging and transport, as well as log-floating techniques.[7] Uittoteho, on the other hand, concentrated on topics related to log floating in northern Finland. In addition to these, the mid-1950s saw the establishment of a committee (Pienpuualan toimikunta) to promote the utilization of small trees. This it did by awarding grants to institutions and researchers for work on technology related to the procurement and use of small trees (Aro, 1958). Milestones in the development of the TTS Institute's department of forestry included the logging championships arranged in 1945, 1951 and 1954. The events provided a setting for new products to make a breakthrough in front of a big public. This was the case, for example, in 1954 when Finnish steel bow saws outdid foreign competitors (Kantola, 1958).

After World War II, a large number of new research organizations was created in the field of forest-technological research, both public and private. The common denominator for all of these new units was that they concentrated on short-term and practical research.[8] Together with motorization, the rationalization of forest work was the main focus of research during the first decades after the World War II. However, still in the late 1960s, there was a big debate in technical journals about the role of horses vs tractors in forest-based work. Fortunately, from a technological development point of view, the Finnish manufacturers of forest machines were continuously developing new and more effective machines for forest workers.

Agriculture

In the early 1950s agriculture still played an important, although a constantly diminishing, part in the Finnish national economy due to the country's late start in industrialization (Pihkala, 1982). It provided livelihoods to approximately one third of the population, which was a considerably larger share than that in some of the more developed countries. In addition to a decrease in farm size, post-war agriculture was characterized by a rapid increase in the use of machines. The tractor, an increasingly common sight on farms in the 1950s, gave a considerable boost to farm productivity and created markets for Finnish manufacturers of tractors and accessories (Pipping, 1955; Pihkala, 1982). While agriculture is not, strictly speaking, a research-based industry (compared to, e.g., the electrotechnical industry), research activities in the field can be traced back all the way to 1881.[9] Wide-scale research, however, did not get fully under way until after the war years.

Research on farm machinery

Research on farm machinery started in 1946 with the establishment of a state-run research institute for agricultural machinery. Its goal was to conduct R&D work on machines, tools and devices used in the farming industry (broadly defined). Prior to this, research on farm machinery, or more exactly the testing of machinery, had been the authorities' responsibility (Kallioniemi, 2002). Activities in the field were boosted in 1953 by a research foundation for agricultural machinery (Maatalouskoneiden tutkimussäätiö), founded by Maatalousseurojen Keskusliitto (Central Union of Agricultural Societies), Maataloustuottajain Keskusliitto (Central Union of Agricultural Producers), as well as Suomen Metalliteollisuusyhdistys (Association of Finnish Metal Industries). The new foundation was much needed because the scarce financial resources of the state-run research institute for agricultural machinery prevented it from taking on many of the test assignments sent to it (Kallioniemi, 2002). In addition to testing, the foundation's duties included the preparation of various types of reports. The Finnish environment and the structure of the country's farming industry posed many difficulties and special requirements to agricultural production and the automation of production. Automation, for example, had to take into consideration the small size and versatile production of farms, the unfavourable growing season, soil types and proportions, forest work and the use of machines in wintertime (Reinikainen, 1958; see also Pipping, 1955). These were some of the topics that testing and research activities focused on.[10]

Research on agriculture, forestry and home economics

The household appliances industry started in Finland as a sideline to the production related to reparations, as the metal industry began to manufacture refrigerators, freezers and washing machines to ease the work of housewives (Pantzar, 2000; see also, Marjomaa, 1979). The beginning of systematic research in agriculture, forestry and home economics can be dated to 1947, when the TTS Institute rented the Matku estate in Nurmijärvi from Oy Alkoholiliike Ab for 50 years. The TTS Institute originally planned to use the estate for its research and training activities but could not launch training in earnest until it received appropriations from the government in 1954. Before this, training consisted mainly of courses dealing with the technical and financial issues related to the rationalization of agriculture, forestry and home economics. The appropriation enabled the Institute to expand its activities into specialist and professional training for the farming industry, as well as continued education for teachers and advisers (Uotila, 1955).

The estate also offered good surroundings for research on field cultivation, farm animals, forestry and home economics. A more extensive research project, launched in 1953, focused on cowsheds. Based on practical methods, the project studied the design of cowsheds and topics related to the comfort of animals (Karhunen *et al.*, 1979). Research on field cultivation dealt with the automation and work methods used for hay and silage. A number of rationalization studies concerning, for example, the growing and harvesting of sugar beet, possible uses for mower loaders and the development of harvesting methods for straw were an important part of operations on the estate. The TTS Institute's goal was to develop Matku into a centralized research centre to house all the various farming industry research projects instead of carrying out individual projects all around Finland (Uotila, 1955).

Research on snowploughs

Research on snowploughs started in 1943, based on the proposal of Professor Lönnroth, then director of the agency for road and water construction, and a donation made by Ahjo Oy's Managing Director Hagert. The initiative was supported by both users and manufacturers of snowploughs, leading to the establishment of a snowplough committee the same year. Representatives of the state, municipalities and various industries, as well as experts from snowplough manufacturers, were invited to participate in the committee's work. The committee was set up to deal with questions concerning the development of snowploughs and winter maintenance.

An essential part of the committee's activities consisted of ploughing experiments that started in different parts of Finland in 1944. These included testing new methods and innovations related to winter maintenance, such as V-type snowploughs and different wing forms, as well as side ploughs and ploughs for snow banks. The structure of ploughs was also investigated and developed for winter conditions in Finland. As a result of extensive research, some plough components were even standardized. In addition to snowploughs, research focused on road gritting, different types of snow and their impact on ploughing. International activities included, for example, participation in the meetings and winter seminars arranged by the Nordic Road Association.[11] In the early 1950s the snowplough committee initiated annual contests to reward the best ideas and inventions in the field. The committee had a total of FIM 100,000 (approximately 16,800 euros) at its disposal for this purpose. Overall, the research expenses spent on the vehicle industry in 1961 accounted for a mere 0.27 per cent of the added value in the industry (Elfvengren, 1963).

Construction industry

Industrialization and population growth maintained strong growth in housing construction up to the war years. During the post-war depression, construction activities faced tough times trying to hire workforce and procure construction material. Although the need for new buildings was great, the shortage of bricks, in particular, slowed down housing construction all over Finland (Simola, 1947). It was not until 1950 that the industry began to pick up thanks to improved material and funding conditions (Tuompo, 1951). The deregulation of construction activities in late 1949 also had a positive impact on growth in the construction industry (Salmensaari, 1951). The upward trend in construction business also led to greater attention being paid to Finnish architecture, exemplified by buildings designed by the likes of Alvar Aalto.

Research on construction

The construction business was booming in the 1950s, attracting some two-thirds of the annual gross investments in Finland. Research on construction also got more active worldwide. Previously focused on construction materials and the technical development of structures, research now came to cover nearly all activities related to the construction industry. Despite the upswing, Finland spent less than 0.1 per cent of all of the funds allocated to the construction business on research activities. Examples of the organizations contributing to this 0.1 per cent include the standardization department of SAFA (the Finnish Association of Architects) and the construction department of Rastori, both of which conducted research to benefit standardization and enhance productivity. Appropriations were scarce and their continuity uncertain, in addition to which research activities were, in general, relatively unsystematic and fragmented (Jarle, 1956).

To achieve better results faster, the Nordic Construction Research Meeting held in 1959 decided that the Nordic countries should coordinate their research in the field. The RAKEVA foundation was set up to manage these issues on Finland's behalf. In addition to coordination, the foundation was also responsible for setting objectives and promoting the financial prerequisites for research in Finland. The foundation's broad-based management included representatives from Finland's highest authorities in construction and research (including three Ministries), top research and education institutes in technology, as well as financial, industrial, technological and professional organizations and institutions with a direct influence on the construction sector (Leiponen, 1961).

Research in the brick industry

In 1948 the Finnish Brick Industry Association decided to found its own laboratory to promote scientific and technical research in the field. The decision was based on the needs presented by extensive post-war reconstruction projects and new tendencies in construction technology. The nature of the brick industry – mainly small scale in Finland – also made it difficult for companies to establish their own factories. The brick laboratory launched operations two years later, in spring 1950, with the task to carry out material testing and raw material research for factories. The laboratory's duties also included brick studies aiming to lay down modern brick standards (Paloheimo *et al.*, 1980). Starting in the early 1950s, the laboratory engaged in mineralogical and technological studies of Finnish clay soils using thermal analysis (Elfvengren, 1963).

From the very beginning, the brick laboratory conducted research in cooperation with international, mainly Nordic, organizations. Practical cooperation between Nordic brick industry associations had started in 1946 in Nordens Samverkande Tegelföreningar (NST), an organization that aimed to solve technological problems common to all Nordic countries. Through the Finnish Ceramic Society, the brick laboratory had close contacts with member organizations and laboratories of the European Ceramic Society (Mäntynen, 1951). Active international operations kept Finnish players up to date on development and innovations in the field. Despite the international nature of the brick industry, R&D activities were carried out on a relatively small scale in Finland. The expenses that the stone, clay and glass industry (which includes the brick industry) allocated to research in 1961 accounted for 0.44 per cent of the added value in the industry (Elfvengren, 1963).

Research in the textile industry

Similar to the sectors discussed previously, the textile industry was not a newcomer in 20th-century Finland. Along with agriculture and forestry, the textile industry was one of the sectors that was not based on research. Attempts to benefit from research were made later, mainly in the form of quality control.[12] Improvements and changes made to products were usually of a minor kind. The industry produced a great number of different products, most of which called for special machines and a staff with special training (Turunen, 1956; see also Lindberg, 1963). The textile industry was characterized by development consisting of small changes to different products. As in most other businesses, systematic R&D was usually launched to solve a specific problem in the field. In addition, changes in fashion and customer needs presented R&D with continuous new challenges (Kestilä, 1965; Jääskeläinen, 1965). The creation of a harmonized measurement system also called for investigation (Ärölä, 1956).

The Finnish Wool Industry Research Laboratory dates back to September 1947, when the Boards of the Wool Manufacturers' Association and the Wool Factory Purchasing Cooperative decided to set up a laboratory to deal with technical questions that had arisen in the wool industry. These included topics related to dyeing and finishing, which the laboratory set out to study in relatively modest conditions. The main reason for establishing the laboratory was the rapid technological development abroad, which the Finnish wool industry wanted to keep up with. Research was considered to be the only way to secure the wool industry's competitiveness. From the very beginning the laboratory's task was defined to be applied research that targeted certain practical problems (Silén, 1949a). This meant leaving basic research in the field to VTT Technical Research Centre of Finland, where they were conducted by deputy research engineer J. Saarinen (VTT, 1949).

The laboratory started up in March 1948 in the Pohjois-Esplanadi office in central Helsinki, focusing on assignments and issues given to it by the wool industry technical committee. The laboratory's first task was to look into questions concerning the removing of mineral oil-based spinning oil. Research was hampered by the poor availability of fatty acids, which forced the laboratory to find alternative methods. The answer was found in Leeds, from studies carried out by Professor Speakman, whose results were successfully applied by the laboratory. The laboratory also studied detergents and found that the quality of Finnish pine soap was on a par with that of modern synthetic detergents. The creasing of fabrics, suitability of dyeing agents and identification of alkali damage in wool were other essential topics in the laboratory's work (Silén, 1949b).

In 1954, the Textile Research Association was founded to conduct research for the textile industry. Although working with limited resources, the Association carried out praiseworthy research in its field and participated, partly with foreign appropriations, in an international research programme on problems faced by the wool industry. Its activities being internationally oriented, the Association did not play a significant role in solving technological problems faced by the Finnish industry (Auterinen, 1965). Despite the activities of the research institutions described above, the Finnish textile industry spent only 0.07 per cent of its added value on research in 1961. In other words, very little research was carried out in the textile industry at the beginning of the 1960s (Elfvengren, 1963).

Company level: input

In many industries, raw and other materials, different stages of manufacturing process, as well as completed products, require a continuous examination of quality. The only place for work such as this was the laboratory. The need for continuous examination of quality was a starting point for laboratories in a large number of Finnish firms. In addition to the customary quality control, the pre-war years was a period during which the need for more demanding types of research activities started to appear. However, it was soon realized that in order to conduct research activities, new tools as well as new personnel with proper skills were needed in laboratories.

According to Freeman and Soete (1997), the distinctive feature in modern industrial R&D is its scale, its scientific content and the extent of its professional specialization. The first indicator of the systematic efforts directed to R&D-type activities was the establishment of company laboratories. The laboratories varied largely in terms of size and activities performed. However, a common denominator for all of them was that they took care of the product development activities of the companies. In addition, people who worked there were concentrated on R&D activities, and had not to take care of any other business. In a large number of companies, laboratories were separated from the other activities of the firm, and R&D personnel often worked in a separate department. This is why there emerged rumours and stories about R&D personnel, saying that they were strange beings with limited communication skills. Even as late as in the 1960s, various technical magazines described them as some sort of.

The first laboratories established by Finnish companies originated from the first decade of the 20th century. Since then, the number of laboratories has continuously increased. In addition, laboratories have been distributed over all the branches of industry, indicating that the developmental activities were widely accepted among the company managers. The laboratories varied largely in terms of size and activities performed. Table 3.2 reports the year of the establishment of separate R&D departments in 61 Finnish companies before 1970, over two-thirds of which were established before 1960. The 1920s clearly saw a breakthrough in the establishment of R&D departments in nine companies.

R&D in companies has witnessed several changes over the studied period. In the beginning of the 20th century, mainly basic research was carried out. It focused, for example, on analyses of the behaviour of new materials under certain circumstances, as well as the control of the quality of products. In the 1940s, a more goal-oriented type of research became common. In order to find out new solutions for existing practical problems, laboratories were established across most branches of industry. While the role of basic research has decreased over the years, except in a few science-based branches of industry, such as chemicals, applied research has become more important. In applied research, the focus is directed primarily towards a specific practical aim or objective (OECD, 1993). Basic research has in the main been left to universities and public research centres.

One can conclude that quite a few research laboratories were established already before WWII and that the period from the early 1940s to the mid 1960s witnessed the emergence of R&D as an important part of company strategy. Despite the minor numbers of personnel involved in laboratories or the low shares of R&D expenditures of the turn-over of the company, laboratories succeeded to establish their places as a momentous part of the company. What was also for the great help was the large number of articles in technical magazines, in which the importance of R&D was discussed as a key to success of a 'modern' industrial firm.

Company level: output

If R&D expenditures can be seen as an input considering scientific-technical activities, innovations represent the output side of the process. It seems fair to say that Schumpeterian economics has functioned as the main source of inspiration for innovation measurement and quantitative empirical research in the field, not least within the OECD. Schumpeter was primarily interested in business cycles and the underlying dynamic processes of the emergence, development and decline of industries, rather than the nature and diffusion of specific innovations in specific industries. He conceptualized the microeconomics of industrial renewal

39

Table 3.2 R&D in some Finnish companies before 1970

Company (year of establishment of R&D unit)	Field of research	No. of staff (year) or R&D/ turn-over	Firm alive in 2000
Tampereen Rohdoskauppa Oy (1907)	Chemical		No
Alko Oy (1908)	Beverage		Yes
Fiskars Oy (1909)	Plastic	0.62% (1966)	Yes
SOK (1913)	Machines		Yes
OVAKO (1915)	Steel	20 (1970)	Yes
Valio Oy (1916)	Chemical		Yes
Paraisten Kalkkivuori Oy (1917)	Mining	1.2% (1965)	Yes
Puolustuslaitos	Chemical		Yes
Lokomo Oy (1920s)	Machinery		No
Kymin Oy (1920s)	Pulp & paper	56 (1968)	Yes
Kemira Oy (1920s)	Chemical		Yes
G.A. Serlachius (1923)	Energy		Yes
Keskusosuusliike OTK (1923)	Machines		Yes
Medica Oy (1924)	Chemical	~100 (1960)	No
Rikkihappo Oy (1927)	Chemical	~1% (1967)	No
Kastor Oy (1928)	Machines		No
Valtion Lentokonetehdas (1920s)	Machines		No
Vuoksenniska Oy (1930)	Metallurgy	0.6% (1967)	Yes
Suomen Autoteollisuus Oy (1931)	Vehicular		Yes
Airam Oy (1935)	Illumination		Yes
Suunto Oy (1930s)	Electrical		Yes
Outokumpu Oy (1930s)	Metallurgy		Yes
Orion Oy (1930s)	Chemical	~60 (1951)	Yes
Enso-Gutzeit Oy (1930s)	Pulp & paper	65 (1970)	Yes
Strömberg Oy (1943)	Electricity	240 (1968)	Yes
Mittari Oy (1944)	Electrical		No
Leiras Oy (1946)	Chemical		Yes
Lemminkäinen Oy (1948)	Asphalt		Yes
Suomen Kaapelitehdas (1949)	Chemical	>100 (1963)	Yes
Finlayson Oy (1940s)	Clothing		Yes
Työase Oy (1940s)	Tools		No
Suomen Malmi Oy (1940s)	Metallurgy		No
Yhtyneet Paperitehtaat Oy (1950s)	Pulp	0.6% (1970)	Yes
Huber Oy (1950s)			Yes
Kotivara Oy (1950)	Foodstuff		No
Teknos-Maalit Oy (1951)	Chemical		Yes
Typpi Oy (1952)	Chemical	2% (1967)	No
Raision Tehtaat Oy (1952)	Foodstuffs		Yes

(Continued)

Table 3.2 Continued

Company (year of establishment of R&D unit)	Field of research	No. of staff (year) or R&D/ turn-over	Firm alive in 2000
Wärtsilä Oy (1955)	Engines		Yes
Berner Oy (1950s)	Chemical		Yes
Tampella Oy (1960)	Paper		Yes
Nokia Oy (1960)	Rubber		Yes
Kestilän Pukimo Oy (1960)	Clothing		Yes
Kaukas Oy (1960)	Clothing	83 (1961)	No
Rauma-Repola Oy (1960)	Pulp & paper	39 (1961)	Yes
Neste Oy (1961)	Chemical	~40 (1965)	Yes
Medipolar Oy (1961)	Chemical		No
Valkoinen Risti Oy (1962)	Chemical		No
Rautaruukki Oy (1964)	Metallurgy	>100 (1967)	Yes
Asko-Upo Oy (1965)	Furniture		Yes
Kone Oy (1966)	Elevators		Yes
Antti-Teollisuus Oy (1967)	Machinery		Yes
Satoturve Oy (1967)	Chemical		No
Enso-Valmet Oy (1967)	Machinery		Yes
Suomen Sokeri Oy (1960s)	Chemical		Yes

Sources: See Saarinen, 2005.

by identifying the subjects and the objects of that process. The objects in Schumpeter's framework are innovations, while the subjects are the entrepreneurs who introduce these to the market in order to gain a monopoly profit (Schumpeter, 1911; 1942; Canter and Hanusch, 1994).

The definition of an innovation, which has been used in this study, relies loosely on the definitions provided in the Oslo Manual (OECD, 1992). An innovation has been defined as an invention that has been commercialized in the market by a business firm or the equivalent. As a minimum requirement, the innovation had to pass successfully through the development and prototype phase, involving at least one major market transaction. The bottom line for inclusion of an innovation in the databases has been, thus, 'a technologically new or significantly enhanced product compared with the firm's previous products'. Only innovations, which have been commercialized by firms registered in Finland have been included.

The innovation data was compiled using a combination of two different methodologies for the identification of innovations: reviews of trade and technical journals, and reviews of the company histories of 46 large Finnish companies. The literature reviews were undertaken during the years 2001–2. First, the population of journals that were eligible for innovation detection was defined. Journals were considered eligible if they were independently edited and regularly published; that is, mere product listings or announcements, irregular publications or journals directly controlled by companies were not considered eligible. This approach resulted in a population of 42 trade or technical journals. In the next phase all such journals were selected that regularly published edited and non-paid material about innovations. The focus was on articles dealing with the introduction of new products which conformed to the definitions and criteria for an innovation. Listings of new products were avoided. Instead, more emphasis was paid on the editorial content of the journals. However, as it turned out, the number of journals was rather limited and not all journals, particularly during the first decades, contained information on innovations. After the selection process of relevant journals, the final list included 36 journals out of 42. Together, the selected journals ensure a proper sectoral coverage of the industrial life in Finland during this particular period.

This type of approach is in contrast to, for example, the work of Kleinknecht and Bain (1993) or the OECD Oslo manual's (1992) guidelines for literature-based innovation collection that also consider eligible non-edited product announcements. However, this latter kind of approach contains a serious risk of high selection bias, as firms have incentives to announce even incremental design modifications, product differentiations and imitations that should not be considered as innovations. On the other hand, trusting professional journal editors should reduce this risk substantially.

In the studies conducted by the Futures Group in the USA (based on the SBIDB-data) (Edwards and Gordon, 1984) and by Acs and Audretsch (1990), a common concern was the representativeness of large firms' innovations. In the study by the Futures Group, the over-representation of the large firms was seen as problematic: 'the material appearing in the new-product sections of the trade journals should only be weighted to the large firm to the extent that the small firm is not sophisticated enough, or does not have the necessary resources, to produce press releases' (Edwards and Gordon, 1984, pp. 14–15).

Learning lessons from the previous exercises, I decided to use annual reviews of large companies for the database. Unfortunately, the older reviews did not contain any information on new products or R&D activities. Instead, the economic development of the firm and market situation

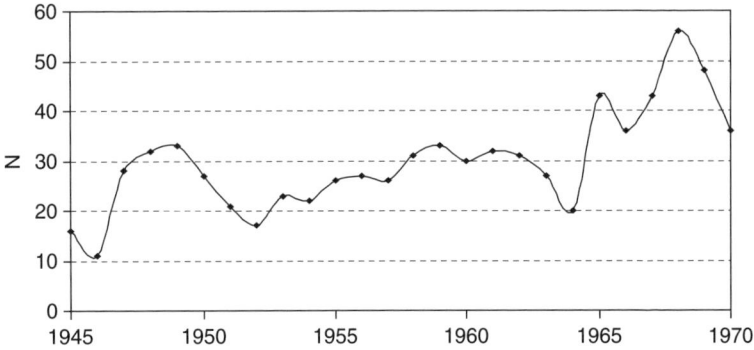

Figure 3.2   Number of innovations according to the year of commercialization

was well described. Information on products and production methods appeared for the first time in the early 1980s, which was close to the final year in my historical time period. Therefore, I decided to include company histories of 46 large Finnish companies as a complement for the annual reviews.

In order to get some idea about the coverage of the innovation data, some basic results are presented. Among all the variables collected, the year of commercialization is probably the best to be included in this study, because this information is available for all of the innovations. It also gives some indications about the long-term development of the innovative pattern in Finland, in rather general terms. Figure 3.2 shows the number of innovations according to the year of commercialization.

The first observation from Figure 3.2 is that the general trend in the number of innovations is increasing over time. After some steady decades in the 1940s and 1950s, the number of innovations achieved an unstable and accelerating pattern towards the end of the studied period. During the period 1945–70, the number of innovations commercialized yearly varied basically (with few exceptions) between 20 and 50 innovations per year. The detected pattern in the figure gives some clear indications about the increasing level of innovative activities of Finnish firms, but also about the fact that the method for identifying innovations is consistent from year to year.

After World War II, mechanical engineering and the forest-based industries constituted the major pillars in the Finnish industrial life. Since the late 1960s, the Finnish science and technology policy has consistently been oriented towards upgrading the knowledge base of the country and to put a great emphasis on the significance of innovation and high technology. As a result, the electronics and electrical industries have entered the club of major sectors.

Conclusion

As shown by these industry-specific reviews, the scientific and technical research carried out in Finland prior to 1970 (excluding the activities of VTT Technical Research Centre of Finland and KCL) was very practical and applied in nature. Research was often instigated by the industry, which needed systematic studies to solve specific problems or to find answers to future development and challenges. Although the industry was, in many cases, behind activities, research organizations were often successful in getting the state to finance projects. The international nature of research and the savings to the national economy that resulted from solving a problem were hardly an obstruction to the granting of appropriations. However, in terms of the sums spent on research, Finland could still be classified as a developing country in the mid-1960s, compared to other industrialized countries.

That research was useful to the future of industry had become a well-known fact in the late 1940s and early 1950s. Nevertheless, both the state and industry showed a clear lack of insight into the matter. In the early 1950s Finland was one of the few countries where no agreement on technical research had been made between the state and industry. The state believed that any expenses incurred from such research should be covered by industry, while the latter felt this was the state's responsibility. This led to a situation in which virtually no one worked to improve matters.

A general observation made after collecting the data for this study was that innovation is not a new phenomenon – neither is systematic R&D activity or a large number of other features that have just recently appeared and been conceptualized in the innovation literature. The first evidence of R&D activities among private Finnish companies can be found in the first decades of the 20th century. Although the R&D departments were then called laboratories, the work carried out there was basically the same as the R&D work performed by companies today. In line with the increase of laboratories and research activities performed by firms, the results of this work can also be seen. The outputs used in this study are product innovations, which have experienced a slight increase in general trend between the years 1945 and 1970.

Notes

1. The history of VTT is discussed in Michelsen, 1993.
2. The early days of KCL are discussed in Jensen's well-written article from 1966, pp. 453–42.
3. For a more comprehensive picture of the research centres in Finland, see Kohonen's table in *Suomen Tieteen historia* 4 (2002), pp. 170–1.
4. As an interesting side note, Finland was ahead of other Nordic countries in the organization and efficiency of R&D in these days. See Levón, 1958, p. 83.
5. (Unknown author) *Puumies* 1963, p. 295. The article is based on a speech by V. J. Palosuo at a celebratory forest event in Salla on 10 August 1963, under the heading 'The Significance of Research to the Competitive Edge of Forestry'.
6. (Unknown author) *Suomen Paperi- ja Puutavaralehti*, 1949, pp. 201–11.
7. (Unknown author) *Suomen Paperi- ja Puutavaralehti*, 1946, pp. 356–7.
8. (Unknown), *Puutavaralehti*, 1949.
9. The early days of R&D in agriculture are discussed in the history of the Agricultural Chemistry Department written by Hirsjärvi. Hirsjärvi, 1957, pp. 35–42.
10. In the mid-1960s, Lokomo Oy initiated extensive development of forestry machinery in cooperation with the Beloit Group from the USA. See Virkkunen, 1967, p. 35.
11. (Unknown author) *Tielehti*, 1958, pp. 29–31.
12. (Unknown author) *Teknillinen Aikakauslehti*, 1960, p. 119.

References

Acs, Z. and Audretsch D. (1988) 'Innovation in Large and Small Firms: An Empirical Analysis', *American Economic Review*, Vol. 78, No. 4, September, pp. 678–90.

Acs, Z. and Audretsch D. (1990) *Innovation and Small Firms*, Cambridge, MA: MIT Press.

Aro, Paavo (1958) 'Metsäteknologisesta tutkimuksestamme', *Paperi ja Puu*, No. 5.

Ärölä, A. (1956) 'Vateva – vaatetusteollisuuden keskuselin', *Teollisuus-Sanomat*, No. 6.

Auterinen, Martti (1965) 'Keskitetyn tutkimustoiminnan tarve ja ajankohtaisuus tekstiiliteollisuudessa', *Tekstiililehti*, No. 5.

Bruland, K. and Mowery, D. (2004) 'Innovation Through Time', in J. Fagerberg, D. Mowery and R. Nelson (eds), *The Oxford Handbook of Innovation*, Oxford: Oxford University Press.

Canter, U. and Hanucsh, H. (1994) 'Schumpeter, Joseph Alois', in Geoffrey M. Hodgson, Warren J. Samuels and Marc Tool (eds), *The Elgar Companion to Institutional and Evolutionary Economics, L–Z*, Aldershot: Edward Elgar.

Edwards, K. L. and Gordon, T. J. (1984) *Characterisation of Innovation Introduced to the US Market*, Report to the US Small Business Administration, Glastonbury: Futures Group.

Elfvengren, Elisabeth (1963) *Teknillis-luonnontieteellinen tutkimus Suomessa vuonna 1961 – Tilastollinen selvitys tutkimushenkilökunnasta ja tutkimuskuluista*, Teknillisten Tieteiden Akatemia, Helsinki: VTT Rotaprintpaino.

Freeman, C. (1987) *Technology Policy and Economic Performance: Lessons from Japan*, London: Pinter.

Freeman, C. and Soete, L. (1997) *The Economics of Industrial Innovation*, 3rd edn, London and Washington: Pinter.

Griliches, Z. (1995) 'R&D and Productivity: Econometric Results and Measurement Issues', in P. Stoneman (ed.), *The Handbook of the Economics of Innovation and Technological Change*, Oxford: Blackwell Scientific.

Hirsjärvi, V. P. (1957) 'Valtion Maatalouskemiallisen laboratorion historiikki', *Suomen Kemistilehti*, 1957.

Hjerppe, Riitta and Vartia, Pentti (1998) 'Talouden kasvu ja rakennemuutos 1860–1997', in H. Loikkanen, J. Pekkarinen, S.-A.Siimes and P. Vartia (eds), *Kansantaloutemme: rakenteet ja muutos*, Helsinki: Taloustieto Oy.

Jarle, P.-O. (1956) 'Rakennusteknillinen tutkimustoiminta', *Teknillinen Aikakauslehti*, 1956.

Jensen, Waldemar (1966) 'The Finnish Pulp and Paper Research Institute: Retrospect and Future', *Paperi ja Puu*, No. 9, 435–42.

Johnson, Ellis A. and Striner, Herbert E. (1960) *Research and Development, Resources Allocation, and Economic Growth*, Operations Research Office, Johns Hopkins University, July.

Jussila, E. A. (1952) 'Puutekniikan tutkimuksen Kannatusyhdistys ry:n synty ja alkuvaiheet', *Paperi ja Puu*, No. 4.

Jääskeläinen, U. I. (1965) 'Eräitä näkökohtia vaatetusteollisuuden tuotesuunnittelusta', *Tekstiililehti*, No. 5.

Kallioniemi, Marja (toim.) (2002) *Sata vuotta tutkittua maataloustekniikkaa*, Maa-ja elintarviketalouden tutkimuskeskus, MTT:n selvityksiä 18.

Kantola, Mikko (1958) 'Metsätyövälineet tutkimuksen kohteena', *Puumies*, No. 9.

Karhunen, J., Pyykkönen, M., Mykkänen, U., Nieminen, L. and Saloniemi, H. (1979) 'Pihattotutkimus 1976 ...1978', VAKOLAn tiedote 29.

Kestilä, E. H. (1965) 'Tuotesuunnittelusta', *Tekstiililehti*, No. 2.

Kleinknecht, A. (ed.) (1996) *Determinants of Innovation*, London: Macmillan.

Kleinknecht, A. and Bain, D. (eds), *New Concepts in Innovation Output Measurement*, London: Macmillan.

Kodama, F. (1986) 'Technological Diversification of Japanese Industry', *Science*, 233, pp. 291–6.

Leiponen, Kauko (1961) 'Maamme rakennustutkimusta on tehostettava. Insinöörilehti', Paperi esitettiin XI Pohjoismainen Rakennustutkimuskokous, 5-7.9.1961, Helsinki.

Lemola, Tarmo (2001) *Tiedettä, teknologiaa ja innovaatioita kansakunnan parhaaksi – Katsaus Suomen tiede- ja teknologiapolitiikan lähihistoriaan*. VTT, Group of Technology Studies, Työpapereita 57/2001, VTT, Espoo.

Levón, Martti (1958) 'Mitä teknillistieteellinen tutkimustyö on ja miten se meillä Suomessa on organisoitu', *Paperi ja Puu*, No. 5.

Lindberg, Joel (1963) 'Tekstiiliteollisuus ja tutkimustoiminta', *Tekstiililehti*, No. 3.

Lindberg, Nils J. (1956) 'Metsäteollisuutemme vuosina 1945–55', *Teknillinen Aikakauslehti*, 1956.

Mäntynen, Matti (1951) 'Teknillistieteellinen tutkimustyö tiiliteollisuudessa', *Teollisuuslehti*.

Marjomaa, Tarja (1979) 'Pyykkilaudasta elektroniikkaan', TTS, *Teho*, No. 9.

Michelsen, Karl-Erik (1993) *Valtio, teknologia, tutkimus: VTT ja kansallisen tutkimusjärjestelmän kehitys*, Espoo: Painotuskeskus Oy.

OECD (1986) *OECD Science and Technology Indicators*, Paris: OECD.

OECD (1992) *Oslo Manual*, OECD/GD (92)26.

OECD (1993) *Frascati Manual*, Paris: OECD.

Paloheimo, A., Lepistö, T. and Suonio, J. (1980) *Suomen tiiliteollisuuden historia*, Suomen tiiliteollisuusliitto r.y./Tiilikeskus Oy.

Pantzar, Mika (2000) *Tulevaisuuden koti – Arjen tarpeita keksimässä*, Keuruu: Otavan Kirjapaino Oy.

Pihkala, Erkki (1982) 'Maa- ja metsätalouden uusi asema', in J. Ahvenainen, E. Pihkala and V. Rasila (eds), *Suomen Taloushistoria 2*, Helsinki: Kustannusosakeyhtiö Tammi.

Pipping, Hugo E. (1955) *Suomen talouselämä toisen maailmansodan jälkeen*, Helsinki: Söderström & Co.

Porter, Michael E. (1980) *Competitive Strategy: Techniques for Analyzing Industries and Competitors*, New York: The Free Press.

Reinikainen, A. (1958) 'Maatalouskoneiden tutkimustoiminta ja koneille asetettavista vaatimuksista Suomessa', *Teollisuussanomat*.

Roschier, H. (1959) 'Puun kemiallinen perustutkimus ja teknillinen jalostus', *Suomen Kemistilehti*.

Saarinen, J. (2005) 'Innovations and Industrial Performance in Finland 1945–98', *Lund Studies in Economic History*, Vol. 34, Stockholm: Almqvist & Wiksell International.

Salmensaari, L. (1951) 'Rakennustoiminta', *Teknillinen Aikakauslehti*, No. 1.

Schumpeter, J. (1911) *Theorie der wirtschaftlichen entwicklung*, Leipzig: Duncker & Humboldt. English translation, *The Theory of Economic Development*, Harvard, 1934; 8th edn, 1968, Cambridge, MA: Harvard University Press.

Schumpeter, J. (1942) *Capitalism, Socialism and Democracy*, Harper & Brothers, Harper Colophon edition 1975.

Serlachius, Erik R. (1959) 'Pohjoismainen yhteistyö laajenee sellusoosan- ja papareintutkimusen alalla', *Paperi ja Puu*, No. 9.

Silén, Gösta. (1949a) 'Suomen villateollisuuden tutkimuslaboratorio' (Esitelmä pidetty Suomen Tekstiilimiesten Liiton kokouksessa 26.3.1949), *Teollisuuslehti*.

Silén, Gösta. (1949b) 'Finska Ylleindustrins Forskningslaboratorium och forskningsproblem inom ylleindustrin' (Föredrag hållet vid Fackavdelningens för kemi möte 5.4.1949), Tekniska Föreningens i Finland Förhandlingar.

Simola, Emil J. (1947) 'Vuosi 1946', *Teknillinen Aikakauslehti*, No. 1.

State Council (1974) *Yritysten tutkimus- ja kehitystoiminnan edistäminen*, Helsinki: Valtioneuvosto, Komiteanmietintö, p. 126.

Stenholm, Taito (1956) 'Perustava tutkimustyö tukkien käsittelystä ja varastoinnista jalostuslaitoksilla', *Puumies*.

Törnudd, E. (1958) 'Teknillisestä tutkimuspanoksesta meillä ja muualla', *Paperi ja Puu*, No. 5, pp. 269–71.

Tuntematon (Unknown) (1946) 'Metsätehon tutkimustoiminta 1945 ja 1946', *Suomen Paperi- ja Puutavaralehti*, No. 22.

Tuntematon (Unknown) (1948) 'Villatehtaiden yhteinen tutkimuslaboratorio', *Tekstiililehti*.

Tuntematon (Unknown) (1949) 'Research Work in Mechanical and Chemical Wood Technology in Finland', *Suomen Paperi- ja Puutavaralehti*, No. 12.

Tuntematon (Unknown) (1951) 'Tutkimustyötä tekstiiliteollisuudessa', *Teollisuuslehti*.

Tuntematon (Unknown) (1963) 'Tutkimuksen merkitys metsätalouden kilpailukyvylle', *Puumies*.

Tuntematon (Unknown) (1958) 'Puolitoista vuosikymmentä uraauurtavaa lumiaurojen tutkimustyötä Suomessa', *Tielehti*, No. 3.

Tuntematon (Unknown) (1960) 'Tekstiiliteknillistä tutkimustoimintaa', *Teknillinen Aikakauslehti*.

Tuompo, Eino (1951) 'Arkkitehtuuri', *Teknillinen Aikakauslehti*, No. 1.

Turunen, Oiva (1956) 'Vaatetusteollisuudesta meillä ja muualla', *Teollisuus-Sanomat*, No. 6.

Uotila, P. J. (1955) 'Työtehoseuran tutkimus- ja opetuskeskus vihitty tarkoitukseensa',. *Insinöörilehti*.

Valtioneuvosto (1974) *Yritysten tutkimus- ja kehitystoiminnan edistäminen*, Helsinki: Valtioneuvosto, Komiteanmietintö 1974, 126.

Vesterinen, Emil (1952) 'Metsätöiden koneellistamisen tarpeellisuus', *Teollisuustalous*, No. 1.

Virkkunen, Veikko (1967) 'Lokomo Oy ryhtynyt laajaan metsäkoneiden kehitystyöhön', *Puumies*.

VTT (1949) 'Valtion teknillinen tutkimuslaitos – Vuosikertomus (Annual report) 1948', Helsinki.

4
The Existence and Nature of Innovation in Mature Industry: Case Study – Finnish Forest Industry

Pekka Pesonen

Introduction

Innovation has been recognized as the core of renewal and the essential factor for competitiveness of companies, industries and, thus, of societies as well. Nevertheless, the focus and rate of renewal varies among companies as a result of diversity in drivers of innovation, that is the factors supporting or hindering innovation; competitive environment, firm's resources, prevailing technology, firm's competences and willingness to innovate, to name a few. Many of these factors are, at least to some extent, industry dependent, and in particular bound to the existing technological paradigm, which is often industry-specific. Therefore, the diversity of innovation activity of firms is reflected to industry level as well. Especially, focus of innovation activity and the rate of innovation have been found to vary across products and industries and, moreover, across the life cycle of industries (Gort and Klepper, 1982; Utterback and Abernathy, 1975). The rationalization behind this phenomenon is the basic idea that industries evolve after birth through different stages to maturity, and the factors affecting innovation evolve as well.

This chapter addresses the specific question of whether a mature industry can be innovative, and, if so, what the focus of innovation is in the particular stage of the industry life cycle. The starting point of the chapter is the idea that individual innovations are the results of the activity in firms aiming at renewal (and of course at increased profits to separate innovation from invention). Thus, single (technological) innovations illustrate the technological development of an industry; the quantity of significant innovations portrays the rate of renewal; and the type of innovations, on the other hand, the focus of renewal.

An ideal example of a mature industry is the Finnish forest industry. It is an industry that is often portrayed as an example of a non-innovative branch, because of the quite non-complex nature of the end products, the categorization as a low-tech industry, as well as its low level of R&D investment. Thus, it is appealing to examine with this case industry whether the general assumption of lack of innovation in a mature industry holds, and, if not, what type of innovations firms in a mature industry pursue. Is the often generalized assumption of mature, low-tech industries being incapable or unwilling to innovate an all-embracing fact? In addition, the global forest industry is experiencing an era of transformation regarding markets, location of production and innovations; therefore, a better understanding of the innovation activity in the industry is needed in order to develop new strategies for innovation and to encourage renewal. This chapter examines the rate of innovation (number of significant technological innovations) over time, and the type of innovations (product or process innovations) in the Finnish forest industry in relation with the industry's maturity.

As the definition of innovation has broadened over the last decades, there is a need to clarify what it means in this chapter. Innovation is approached from the technical perspective and an innovation here is regarded as one to embody tangible technology. In other words, we speak of technological innovations and exclude the pure service, marketing and organization innovations, which are more intangible innovations and often relate to the way the innovator operates or acts (renewal in activity). Moreover the definition used here follows the more traditional one of the OECD (1997) and incorporates products as being technological innovation. An innovation is considered as being new or a significantly improved product from the innovator's perspective. The definition of innovation supports the aim of the chapter in analyzing the technological renewal of an industry. The study looks at single innovations which are novel compared to the previous technology, and thus constitute the development of predominant technology.

Industry life cycle and innovation activity

When an industry evolves over time its competitive environment changes, as do the characteristics of the organizations active in it. Moreover, it is these changes in the nature of the industry and of the operating firms that define the stage of the industry life cycle. Business surroundings evolve as, for example, new trends in consumer behaviour take place, new laws or regulations are enforced or competing technologies emerge. Organizations, on the other hand, change by acquiring new capabilities and competences, developing their technology and operational skills, or creating new collaboration linkages, for instance. These issues affect companies' actions to develop successful innovations and to gain competitive advantage in various ways (see Palmberg, 2006). At some point in the development curve of an industry, vast innovation activity can be the key to success, while, in another, firms do not pursue innovation at all. Innovation can also vary by type as firms might strive for new market offerings at one point, or renewed in-house processes at another. Innovation and innovation activity of firms in mature industries is emphasized to have a peculiar nature that differs from other stages of industry evolution.

There is no universal definition for what is considered as a mature industry. However, there are some common elements that are often used when describing what is regarded as an industry experiencing the mature phase of industry life cycle. In his profound work of industry life-cycle theory, Klepper (1997) distinguishes three major stages in industry evolution. These stages have similarity with the stages of production process technology proposed by Utterback and Abernathy (1975), by nature of innovation and basis of competition. Three basic stages are: initial, exploratory stage (1); growth stage (2); and mature stage (3).[1] In the first phase, market volume is low, product design is undeveloped and no dominant design exists, while relatively few active firms seek product innovation driven by the market needs. In the growth stage, products begins to stabilize and production technology becomes more specialized and efficient, markets grow as do the number of firms, and shakeouts occur. After growth, entry rate decreases and industry experiences a stabilization in terms of market shares and product innovation, while technological development slows down, gearing towards process improvement.

The comprehension at the present is that maturity, which in many aspects represents pattern of stability, takes place after the more dynamic growth stage. Maturity is considered as being a later stage in industry life cycle that is reached after emergence/initial and growth/development phases, as summarized by Klepper (1997). Mature industry is typically described by characteristics like high capital intensity, high productivity that has improved over time, firms producing more or less standardized products, industry consisting of a reducing number of firms, large process scales existing and market price acting as a dominant competitive factor (Utterback and Abernathy, 1975; Klepper, 1997).

Mature industry is also often linked to a particular pattern in innovation that reflects firms' competition strategies as well as the industry's innovation drivers, like the nature of markets. Previous research on innovation and industry life cycles has cast light on issues of the quantity and focus of innovation in firms. It is portrayed that over time, as an industry matures and the number of firms decreases, innovation activity depletes, creating fewer technological novelties, and the focus on renewal shifts towards processes instead of products (Utterback and Abernathy, 1975; Klepper, 1996).[2]

In this chapter, following the aforementioned views on the issue, mature industry is defined as one experiencing decreasing entry rate of new firms, decreasing or stabilized level of firm population and high level of productivity that has increased over time.

Lowering rate of innovation

As an industry matures, competing firms tend to develop mainly minor improvements, that is, incremental innovations, and the rate of technological development decreases. The phenomenon of technological entrenchment can be explained by issues such as the evolution of the competitive setting towards more of an oligopoly and company homogenization (Utterback and Abernathy, 1975; Klepper, 1996), path dependency and technological maturity of firms (Dosi, 1984; Nelson and Winter, 1982), high competitive power of large incumbents aiming at competence enhancing renewal (Tushman and Anderson, 1986) and lock-in to old routines and technologies (Arthur, 1989). In addition, institutional lock-in can be a factor interacting or reinforcing technological lock-in (Foxon, 2002). Classical examples of technological dis-advancements (lock-in to certain technology regardless of superior choices) are the dominance of QWERTY design over dvorak in typewriters (see Utterback, 1994), and the dominance of VHS over Betamax in video cassettes.

When it comes to the innovators, in contrast to old incumbent firms, entering companies have been detected to present a higher probability of innovating (Huergo and Jaumandreu, 2004).[3] Industry newcomers are often considered more flexible and agile in innovation compared to old actives in the field. Established firms have old frameworks for identifying and applying opportunities, as well as for creating new innovations. Thus, their ability to renew might be a more time-consuming process compared to new firms. For instance, in an emerge of an architectural innovation, an incumbent firm faces problems as it needs to reorient itself from one of refinement within a stable architecture to active search for new solutions within a constantly changing context in order to build new architectural knowledge (Henderson and Clark, 1990). Hence, renewal of the company population of an industry can have a positive effect on the rate of innovations as well, whereas lowering entry rate of companies can hinder innovation.

The reason for a declining rate of innovation can also be inner-organizational, for example, related to innovation management. In a mature industry with an oligopoly, firms are often large, and mature as well. These kinds of firms often experience a lack of sustained innovation as a result of unsolved innovation-to-organization problems (Dougherty and Hardy, 1996). Successful innovators manage to solve a higher proportion of problems on the project level, including acquiring resources, building effective inter-functional teams and conceptualizing their new products more thoroughly. Organizational-level problems reflect poor management and/or non-existent innovation strategy and culture. Sustained innovation in mature manufacturing firms essentially needs supportive and nurturing management. Actions and efforts of individual people will lead to a number of innovations, but they will not have the power to create innovations continuously. Innovation activity requires management in order to support innovation, enhance innovative culture and to build, formulate and implement a successful strategy steering this activity. In other words, individuals create innovations, but organizations 'cultivate' innovation management. A more sustained approach to developing an organization-wide capability for continuous innovation requires both of the above-mentioned aspects. If this is not achieved in mature firms, the rate of innovation in the industry may well decrease.

Focusing on process innovation

The second point stressed in the literature, in addition to the lowering rate of innovation, is the changing focus in innovation activity in a mature industry. In other words, firms are increasingly geared to develop process, rather than product innovations. In a developed industry, that has become an oligopoly, efficiency and economies of scale in production are emphasized (Utterback and Abernathy, 1975). Thus, firms tend to aim at renewing their in-house processes, especially in production. Process-orientation is related to firm size, as the process share of total R&D increases with business unit size (at a declining rate) (Cohen and Klepper, 1996). One main goal of firms is to increase the productivity through process development, whereas products in the technological paradigm have become more or less standardized. Product renewal might not give competitive advantage, at least in the specific technology, as product improvements are costly, opportunities are hard to identify and the risk of failure has increased. The idea of pursuing production efficiency in maturity is supported by the fact that firms focusing on process innovation are more often operating in steady markets that do not grow rapidly (Capon *et al.*, 1992). There might be no room for new product offerings, but efficiency in the production of the existing products is seen as the most desired innovation strategy. Major innovations have also been noted to play an important role for growth in productivity (Geroski, 1989).[4]

However, as the prominent share of literature points towards a linkage between maturity and a low rate of significant innovation, McGahan and Silverman (2001)[5] found systematic evidence that innovation is not significantly lower in mature industries than in emerging industries. They also found evidence contradicting the assertion that in the mature stage of an industry innovation is more process- than product-oriented. Accordingly, the mainstream view of mature industries experiencing a lowering degree of technological advancement, where the focal point is in processes, might not be an all-embracing phenomenon. Moreover, there seem to be few studies on innovation in mature industry applying single innovations as the output of innovation activity and as the object of study. This study, aims to cast new light on these issues.

Case industry

Finnish forest industry has a long history, dating back to the 16th century and the industrial manufacturing of sawn wood. Industrialization experienced a rapid boost in the 19th century, when sawmill technology advanced and the first paper machines took place in Finland. Since then products have become quite different. Yet, sawn wood, veneer, pulp and paper have been the main products for over a hundred years, and still are. However, the features and the quality of the products have evolved throughout the industrial era, together with the production technology. In other words, for example, paper is still paper, but it is whiter, lighter, durable, more opaque and produced with lower unit costs, roughly speaking.

The structure of the industry has changed throughout the 20th century. Especially during the 1980s and 1990s, the industry experienced radical changes, as firms accelerated their internationalization and mergers followed each other. Nowadays, there are basically three significant concerns (each having a number of subsidiaries), acting worldwide in both pulp and paper, as well as in wood-product industries. In addition, there are, of course, smaller companies, though only a handful in the pulp and paper industry, which mainly operate in domestic markets.

Development in the Finnish forest industry in the 20th century can be characterized by an enormous growth in production and productivity.

Total paper production grew from 2 million tons in 1960, through approximately 7 million tons in 1985, to 14.3 million tons in 2007 (Finnish Forest Industries Federation, 2008). At the same time, there has been a shift in production from sawn wood to paper products. From the societal perspective, the importance of the forest industry has been crucial. The share of forest products of the total value of exports in Finland was some 42 per cent in the 1980 and, at the same time, the value of forest products comprised 6.7 per cent of the GDP. Corresponding figures in 2006 were 20 per cent and 3.8 per cent. As the national dependence on the industry has diminished, the industry has become ever more international, exporting almost 90 per cent of the total domestic pulp and paper production and some 60–70 per cent of the total amount of domestically produced wood products. In addition, some half of firms' production is nowadays overseas.

As mentioned before, the basic technology of the industry has remained the same over the years. Thus, the industry is often regarded as a non-innovative one. Nevertheless, it has to be taken into account that innovation in the industry does not solely come from inside forest industry companies (innovations over the product life cycle often originate outside the set of current producers, as argued by Gort and Klepper, 1982). The focus has been on production and productivity improvement, where the companies in other industrial classes have played a key role. Machine producers, the chemical industry, electronics and automation, to name a few, have had, in collaboration with the forest industry companies, a huge impact on the technological development of the forest industry. Therefore, it is evident that studying innovation and technological development of the forest industry requires a wider perspective, taking into account innovations developed by companies outside the forest industry, but for the use of the forest industry. The present chapter applies this wider perspective in studying the rate and type of innovation in the mature industry over time.

Data and methods

Innovation in this study is defined as a new or significantly improved product from the perspective of the organization behind the development of the innovation, which has then been commercialized by a firm or equivalent. The definition is based on the database used for acquiring the innovation data for the analysis, namely SFINNO innovation database.[6] SFINNO consists of close to 4500 individual innovations commercialized by Finnish companies between 1945 and 2005 (in spring 2008). Data on single innovations include information on the characteristics of the innovation, on the innovation process and on the commercializing firm.

Individual innovations from the database were included in the analysis according to three criteria: 1) innovation was commercialized by a company operating mainly in the forest industry, 2) technology of the innovation relates it to the forest industry, or 3) innovation is used/implemented in the forest industry. In total, 515 innovations fulfil at least one of these requirements and were selected for the study. In general, these innovations represent the technological development of the Finnish forest industry through the lens of single, technologically significant innovations, developed or used in the Finnish forest industry.

As the study focuses on the number of innovations, one possible source of error is the miscalculation of the number of innovations introduced every year. However, as the method for identifying the innovations in SFINNO has remained unchanged over the years, the sample of forest industry innovations is identified in a similar way during this time. Thus, I argue that the numbers of innovations in different years are comparable against each other, and thus it is safe to analyze the changes in the rate of innovation. The same argument holds for analyzing the relation of product and process innovation intensities. However, it is evident that with the object-based method of identifying innovations, as used in SFINNO, the whole population of innovations is not identified. As the database consists of a sample of the economy's innovations, it is likely that yearly variations may exist. To minimize the variation in the count of yearly commercializations of new products, an average of multiple years' innovation count is used to illustrate the rate of innovation.

Distinction between product and process innovation (type of innovation) was made according to the technological class of the innovation. Innovations technologically representing wood products were labelled as mechanical product innovations, and where the technology of innovation represents pulp and paper products, the label chemical product innovation was given. Innovations technologically representing machinery and equipment, yet used in the forest industry, were labelled as process innovations of machinery and equipment. All other innovations were labelled as other process innovations, including, for instance, software innovation used in process control. In other words, most of the process innovations are not developed (mainly) by the forest industry companies (although some are), but implemented in the production processes of these companies. The type of innovation is independent from the company, or its industry, that is behind the development and commercialization of the innovation. Data on industry level (number of firms, number of entering firms, number of exiting firms, productivity) was acquired from Statistics Finland's annual reports on Finnish industries and from statistical databases with the help of experts from the statistical office. Forest industry includes two industry branches: 1) manufacture of wood and of products of wood and cork, except furniture; and 2) manufacture of pulp, paper and paper products.[7] These industrial branches are labelled later on in the chapter as: 1) wood products and 2) pulp and paper.

Results

Let us begin with looking at the industry life cycle and whether the case industry can be regarded as mature. This is done by analyzing the number of new as well as exiting firms, the number of active companies in the industry and industry productivity. The Finnish forest industry consists of two industry branches that are close to each other in terms of raw material, value chains and because of the fact that the largest companies in forest industry act in both branches as multiproduct firms. However, the population of competing firms in these two branches is very unequal and because of this they are illustrated separately in Figures 4.1 and 4.2.

As Figures 4.1 and 4.2 show there has been quite a similar pattern in both of the forest industry branches regarding the entry and exit of companies. The number of new establishments per year has gone down during the last ten years of observations, especially in the wood products industry. In pulp and paper, the number of entries, as well as exits, is lower and yearly variations in these are bigger, than in wood products.

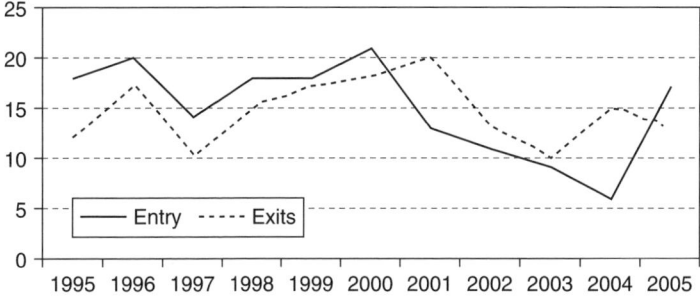

Figure 4.1 Entry and exit in pulp and paper industry. Number of firms by year (Statistics Finland)

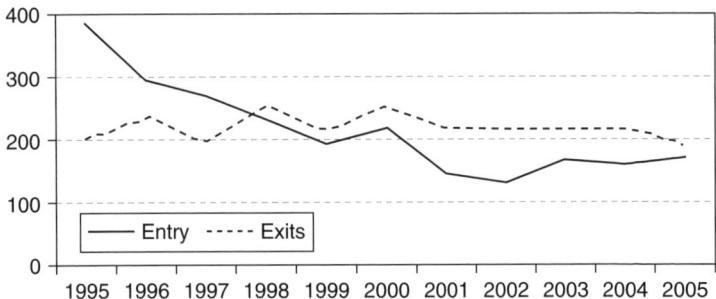

Figure 4.2   Entry and exit in wood products industry. Number of firms by year (Statistics Finland)

However, the trend in both industrial branches has been that the entry rate has gone down, which signals industry maturity; in wood products this occurred significantly at the end of the 1990s and in pulp and paper in the beginning of the 2000s. At the same time exit rates have remained quite stable. In pulp and paper there have been ten to twenty exits every year and in wood products some 200–250. In both cases, the rate of exiting firms has exceeded the rate of entering ones. In wood products this change happened in 1998, when the number of exiting firms exceeded the number of new firms. Since then, there have been more companies exiting the industry branch than entering it. In pulp and paper, the number of exiting firms surpassed the entering one a bit later, in 2001. After that, only in 2005 were there slightly more firms entering than exiting the industry branch. Regarding the entry rates, the Finnish forest industry seems to be mature, even showing a slight decline in the 21st century as more firms exit the competition in the industry than enter it.

The historical analysis of the industry evolution reveals that the firm population in the global pulp and paper industry has decreased throughout the 20th century (Lamberg and Ojala, 2007). Especially in Europe, the number of companies went down between the years 1974 and 2000, halving from 1548 to 763 firms. In Finland the evolution after 1986 shows that the number of active firms rose at first and later on declined (Figure 4.3), as the entry/exit ratios gave a reason to presume they would. The Finnish forest industry was growing at the end of the 1980s and the number of active companies increased quite rapidly, reaching its peak in 1991–2. In consequence of the recession at the turn of 1990, many firms exited the industry and the number of actors competing in the industry underwent a downfall in 1992–3. In the mid-1990s, the company population grew again until 1997, after which it has declined in wood

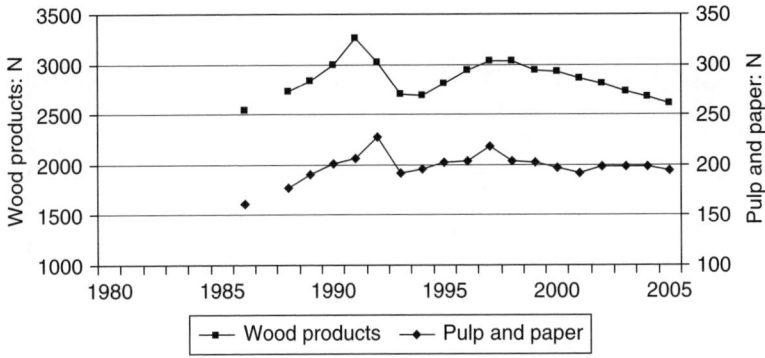

Figure 4.3 Total firm population in the industry. Number of firms by year (Statistics Finland)

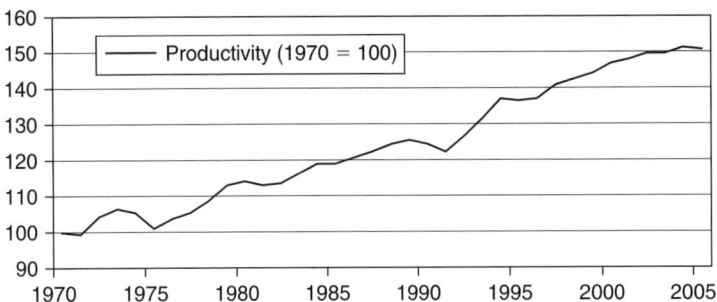

Figure 4.4 Overall productivity index of the Finnish forest industry (Statistics Finland)

products as well as in pulp and paper. In wood products the number of firms came down from 3039 in 1997 to 2609 firms in 2005, and in pulp and paper the fall was from 218 firms in 1998 to 195 firms in 2005.[8]

Industry productivity has increased significantly over the last 35 years. The level of overall productivity (Figure 4.4) shows almost a continuous improvement, resulting in a status in 2005 in which the industry is capable of producing 50 per cent more goods with the same input compared to 1970. However, we can notice a slight decline in the growth of productivity since the end of the 1990s. As a conclusion, it can be stated that the industry is well mature from the viewpoints of entry rate, firm population and productivity. This is what one might expect after what has been noted regarding industry maturity at a global level in the pulp and paper industry (see Lamberg and Ojala, 2007). Next, we will concentrate

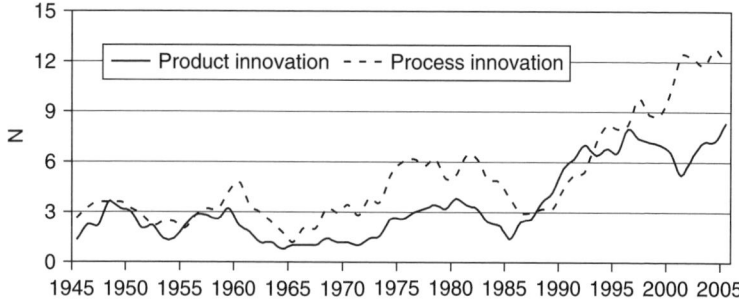

Figure 4.5   Rate of product and process innovation in the Finnish forest industry
(5-years moving average)
Source: SFINNO innovation database.

on analyzing the existence and the nature of innovation in the mature
industry. The rate of innovation in the industry reveals an interesting pattern of
technological development. We can observe three eras of increased innovation which are illustrated as 'waves' in Figure 4.5. This proposes that
there exists certain cyclicality in innovation in the industry, appearing
as times with faster renewal and higher number of innovations,[9] and as
the opposites to this. The *innovation waves* took place in the late 1940s,
in the late 1950s and around the turn of the 1980s. Notable in these
waves is the fact they do not purely consist of either process nor product innovation, but both. This, I believe, is because of the nature of
the particular technology of the industry, and, moreover, because of a
certain duality of innovating in the particular industry; processes and
products are developed in interaction where the development of the one
requires or enables the development of the other. In the late 1980s, the
rate of innovation started to increase again, this time quite rapidly, and
has done so until today. If the pattern repeats itself, we should witness a
stabilizing and after that decreasing rate of innovation in the industry in
the future, forming the fourth innovation wave. Nevertheless, while the
development of products and processes is interconnected and the rate of
new wood products has been relatively high for the past fifteen years, it
might be that process improvements continue to emerge and the peak
of the fourth wave is still to come.

The discovered pattern of innovation waves or cycles contradicts the
common understanding of mature industry in relation to innovation.
First of all, there seem to be no individual trends for product and process

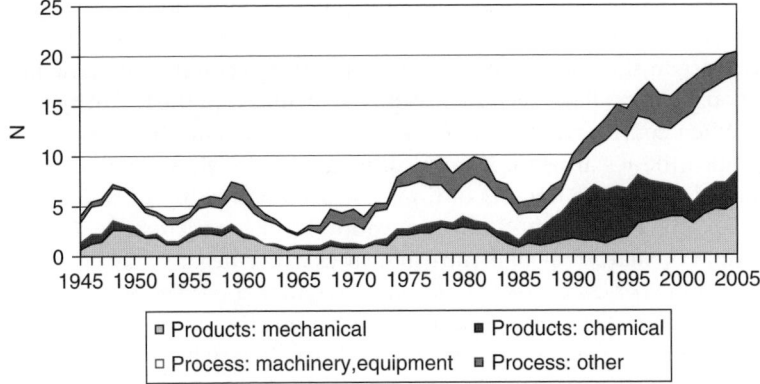

Figure 4.6 Rate of Finnish forest industry innovations, divided by type of innovation (5-years moving average, N = 515)
Source: Pesonen and Saarinen, 2007.

innovation, as Utterback and Abernathy (1975) propose; they seem to evolve in a similar way over time. Second, product innovation does not experience a continuous downfall (after a peak that can be assumed to have taken place around the turn of the 1900s) as industry matures. However, product innovation does grow around the turn of the 1990s, together with increasing firm population (see Figure 4.3) and stabilizes in parallel with lowering/stable entry (see Figures 4.1 and 4.2). Third, process innovation does not increase and then decrease, but it shows a fluctuating pattern. Furthermore, process development experiences a continuous increase since the peak in the firm population in the beginning of the 1990s (Figure 4.3).

The difference between the latest growth in innovation and the three previous waves is the focus inside product innovation (see Figure 4.6). In the fourth, still-growing wave, product innovation has focused on pulp and paper (Products: chemical, ~1987–2000) at first and then shifted to wood products (Products: mechanical, ~1995–2005). However, in the three previous waves the product innovation concerned mainly wood products. This is in line with the changing focus in production, as in the 1980s it shifted more from wood products to pulp and paper. Thus, the technological development has concentrated on the products strategically more important for the firms, which since the 1990s has been paper. It is difficult to say, on the basis of this evidence, whether it is the possibilities in creating new products that come first followed by process improvements, or vice versa.[10] In the fourth wave, however, it seemed to be that product innovations boost process innovations as new pulp and paper products were followed by improvements in processes and the increase in the rate of process innovation might continue because new wood products have been increasingly introduced in the recent years.

When considering the maturity of the industry and the rate of innovation, the results do not fully coincide with previous studies in this matter. While the number of firms has been decreasing since the mid-1990s, the rate of significant innovation has been increasing. Thus, it cannot be stated that all mature industries with a negative net entry rate of firms experience a reducing pace of renewal appearing as a lowering rate of innovation. However, the focus on process innovation in maturity is certainly evident in the case industry. It may be that while, in a mature industry, opportunities for growth are small and more often relate to processes, they exist more frequently (or at least firms that compete based on efficiency and cost reduction seek them more frequently).[11] The introductions of new innovations follow each other as companies try to capitalize on whatever opportunity is open for them. Every possibility becomes valuable, no matter how small the gained benefit might be. This leads to higher number of process innovations, but not necessarily to lowering number of significant product innovations. These product innovations can very well be radical by nature,[12] especially in relation to the otherwise more or less standardized product environment, and thus they are in the interest of the type of firms seeking bigger competitive advantage. In other words, in the forest industry, continuous process improvement is the main target of technological change, but not at the expense of product renewal that can very well support the emergence of a new technological paradigm.

Between product and process innovation, the focus in the Finnish forest industry has been increasingly on process innovation in the past fifteen years. At the same time, the rate of product innovation has remained quite stable, although exhibiting a shift from pulp and paper to wood products.[13] Greater attention to process renewal in the industry is also pronounced by another source of innovation data, namely the CIS (Community Innovation Survey), which shows that a larger share of companies have developed a process innovation than a product innovation in the recent years.[14] In addition to the process innovations developed by the forest industry companies, many of the changes in production technology originate outside the industry. Hence, the importance of suppliers, like machinery and equipment manufacturers in this case, cannot be underestimated, as their significance for technological development seems to grow as industry matures.

It seems that companies in a mature industry experience a certain lock-in phenomenon to present process technologies. After pursuing cost reduction and economies of scale, factories and production processes in general represent high capital intensity. As a result of this, firms are bound, at least to some extent, to products produced by the dominant process technologies, which they have adopted and concentrated to develop further in maturity. New product opportunities are scarce and firms apply the short-term value creation option by improving the processes rather than seeking new radical products. However, process innovation will produce declining benefits, forcing firms to seek new technological trajectories through various product innovations as well.

Discussion

This chapter analyzed innovation activity in a mature industry, and in particular the rate and type of individual, significant technological innovations. The approach is very much technology-oriented, derived partly from the database used, and partly from a choice to give the study a clear focus. Innovation activity and technological development of the industry in the study is described through individual innovations. The role of market, economic, institutional and social factors are not considered in the chapter. Nevertheless, they are acknowledged to have an effect on technological development, as well as the creation and selection of advancements in a technological trajectory (as argued by Dosi, 1982). These, and other possible drivers of innovation, were not considered in detail as the focal point was simply to study the rate and character of innovations in a mature industry, the Finnish forest industry in this case.

When considering the maturity of the industry and the rate of innovation, the results do not coincide with previous studies of this matter. While the number of firms has been decreasing in the industry (signalling maturity), the rate of significant innovation has been increasing. Thus, it cannot be stated that all mature industries experience a lowering rate of innovation.

The findings related to the case industry in particular are in line with the notion put forward by Utterback (1994). He states that the idea of dominant design does not hold for pulp and paper products (or other non-assembled products), but the technological effort is concentrated on improving the enabling technology, that is, production processes. There was found to be cyclicality in the rate of innovation, signalling new, emerging technological trajectories (or changes from one to another) in the industry, but not necessarily transformative changes. These 'waves' found in the rate of innovation might reflect somewhat similar phenomenon that 'dominant design', but more in the sense that there exists some 'enabler' or 'positive change' at the beginning of each wave that supports the technological development in both products and processes. A relevant question to consider for mature industries is how to unlock from this stagnant stage. What should be done in order to get back on the growth stage and increasing returns? It is often stated that radical changes in technology or in market needs related to *existing* products can help companies to 'de-mature' (see, e.g., Dowdy and Nikolchev, 1986). However, I suggest that renewing out of maturity can also happen by creating totally *new* products areas. Identifying the possibility and developing a new product life cycle can form a growing business for mature companies. In the case of forest industry, escape from maturity can happen through the emergence of new product segments. This could mean, for instance, producing biofuels out of wood.

Regarding the implications for future innovation studies, it might be valuable to raise the question of maturity in relation to the product segment level and not industry level. This is because one industry can entail several different products, which can themselves experience different stages of life cycle. Therefore, the implications of these results are twofold. First, the maturity of an industry depends partly on the maturity of its products. One product can be on the mature side, while another product is still experiencing differentiation and can represent the growth stage. Second, and related to the previous point, focusing on product life cycle rather than industry life cycle might reveal a more relevant relationship between technological maturity and the nature of innovation.

Analyzing innovation activity in relation to industry life cycles might not be valid across sectors. Due to differences between industries, and especially between their technologies and business environments, maturity and innovation do not interact similarly in various industrial sectors. There can be multiple technological trajectories in one industry which make the topic at hand more tied to product life cycle than industry life cycle. The results of the chapter show that Finnish forest industry is one of the cases that do not fit the typical theory of innovation and industry life cycle. Thus, it might be more plausible to distinguish industries by their innovation activity and innovations than stage of life cycle, as also inferred by McGahan and Silverman (2001).

Limitations

Obviously, the main limitation of the chapter is the analysis of a single industry, Finnish forest industry in this case. Thus, the results cannot be unambiguously generalized in a wider perspective. However, the chapter certainly indicates a need for cross-sectional study on innovation and development of industries based on single innovations.

Notes

1. It should be noted that the terms 'industry life cycle' and 'product life cycle' (PLC) are often used interchangeably, while similar stages are present in both. In this chapter, the first term is used, but the possibility that one industry can have several products with individual life cycles is acknowledged. However, the focus is on the level of an industry, and in this case the industry has only few major product areas that constitute a large share of the markets and their existence and life cycles can be regarded as analogical with the entire industry.

2. These stylized facts are, however, challenged nowadays in other studies as well (see Windrum, 2005).

3. In Huergo and Jaumandreu's (2004) study of manufacturing industries, the phenomenon of higher innovativeness of entering companies was emphasized, particularly in the case of paper and printing products. This is an important finding, especially in relation to this chapter, as the present case industry is close to the above-mentioned one.

4. Though, it should be mentioned that Geroski did not distinguish between product and process innovations, but spoke of major innovations in general.

5. It should be noted that McGahan and Silverman (2001) used patent counts as proxies of innovation activity. This approach has problems as it is evident that not every innovation is patented and not every patent is an innovation. Patent count is also a rough approximation of the count of innovations as there can be a number of patents per single innovation. Thus, using data on actual innovations rather than patents gives, in the author's view, more useful information on innovation activity and technological development.

6. For a deeper clarification regarding the database and the definition of innovation, see the chapter concerning the database in the beginning of the book and Palmberg *et al.*, 1999.

7. SIC (Standard Industrial Classification) 2002 is used to define forest industry (see Statistics Finland: http://www.stat.fi/meta/luokitukset/toimiala/001-2002/d_en.html).

8. Changes in the total number of active companies do not follow exactly the changes in entry and exit rates as these two analyses are not fully comparable, but rather complementary. This is because business operations, such as mergers and split-ups, account differently for entry and exit rates than they do for the number of all active companies. For instance, a merger accounts for one entry and one exit, but decreases the number of firms by one. In addition, number of active companies (Figure 4.3) illustrates the average count of all active companies that have been operating for more than six months during the respective calendar year and exceed the minimum threshold set for turnover or number of employees (to show that the firm, in fact, has business activities).

9. Windrum (2005) found that several innovation cycles occurred also in the camera industry over its lifetime, instead of one cycle of radical product and process innovation. One explanation for the waves can be that they resemble technological trajectories (see Dosi, 1984), where innovation speeds up after the introduction of new trajectory and decreases when the developments related to the specific technology become more difficult to capitalize on.

10. Pesonen (2006) pointed out that it can occur either way, as the production of a new product often requires modifications in the process, and, on the other hand, the improvements in the production process often enhance a certain feature of the product.

11. Increasing concentration in the strategic focus – (i.e. business strategy of differentiation, clear long-term vision and high-quality external communications) can also explain the support of innovation in a mature industry (see Pearson *et al.*, 1989).

12. Recent examples include wood-plastic composites, intelligent packages and wood-based chemicals, for instance to help in cancer treatment.

13. Regarding products, the industry has not transformed significantly in the past few years. Main product segments have stayed unchanged; paper and sawn wood still constitute the most important share of sold goods. Nevertheless, innovation in products has created new segments, though they still have rather small markets. These new products are in many cases defined as hybrid products that are created on overlapping technologies, like, for example, wood-based biofuel.

14. In 2002–4, the share of companies that developed a product innovation was 14.4 per cent in wood products and 29.5 per cent in pulp and paper, and the share of companies that developed a process innovation was 32.0 per cent and 35.1 per cent, respectively (Statistics Finland: results from CIS4).

References

Arthur, W. B. (1989) 'Competing Technologies, Increasing Returns and Lock-in by Historical Events', *Economic Journal*, Vol. 99, No. 394, pp. 116–31.

Capon, N., Farley, J. U., Lehmann, D. R. and Hulbert, J. M. (1992) 'Profiles of Product Innovators among Large US Manufacturers', *Management Science*, Vol. 38, No. 2. pp. 157–69.

Cohen, W. M. and Klepper, S. (1996) 'Firm Size and the Nature of Innovation within Industries: The Case of Process and Product R&D', *Review of Economics and Statistics*, Vol. 78, No. 2. pp. 232–43.

Dosi, G. (1982) 'Technological Paradigms and Technological Trajectories: A Suggested Interpretation of the Determinants and Directions of Technical Change', *Research Policy*, Vol. 11, pp. 147–62.

Dosi, G. (1984) *Technical Change and Industrial Transformation: The Theory and an Application to the Semiconductor Industry*, London: Macmillan.

Dougherty, D. and Hardy, C. (1996) 'Sustained Product Innovation in Large, Mature Organizations: Overcoming Innovation-to-Organization Problems', *Academy of Management Journal*, Vol. 39, No. 5, pp. 1120–53.

Dowdy, W. L. and Nikolchev, J. (1986) 'Can Industries De-Mature? Applying New Technologies to Mature Industries', *Long Range Planning*, Vol. 19, No. 2, pp. 38–49.

Finnish Forest Industries Federation (2008) 'Statistics', accessed in May–November 2008. Available at: http://www.forestindustries.fi/tilastopalvelu/Tilastokuviot/Pages/default.aspx

Foxon, T. J. (2002) 'Technologcial and Institutional "Lock-in" as a Barrier to Sustainable Innovation', ICCEPT Working paper, November.

Geroski, P. A. (1989) 'Entry, Innovation and Productivity Growth', *Review of Economics and Statistics*, Vol. 71, No. 4, pp. 572–8.

Gort, M. and Klepper, S. (1982) 'Time Paths in the Diffusion of Product Innovations', *Economic Journal*, Vol. 92, No. 367, pp. 630–53.

Henderson, R. M. and Clark, K. B. (1990) 'Architectural Innovation: The Reconfiguration of Existing Product Technologies and the Failure of Established Firms', *Administrative Science Quarterly*, Vol. 35, No. 1, pp. 9–30.

Huergo, E. and Jaumandreu, J. (2004) 'How Does Probability of Innovation Change with Firm Age?', *Small Business Economics*, Vol. 22, No. 3. pp. 193–207.

Klepper, S. (1996) 'Entry, Exit, Growth and Innovation Over the Product Life Cycle', *American Economic Review*, Vol. 86, No. 3, pp. 562–83.

Klepper, S. (1997) 'Industry Life Cycles', *Industrial and Corporate Change*, Vol. 6, No.1, pp. 145–81.

Lamberg, J.-A. and Ojala, J. (2007) 'Evolution of Competitive Strategies in Global Forestry Industries: Introduction', in J.-A. Lamberg, J. Näsi, J. Ojala and P. Sajasalo, *The Evolution of Competitive Strategies in Global Forestry Industries: Comparative Perspectives*, Dordrecht: Springer.

McGahan, A. M. and Silverman, B. S. (2001) 'How Does Innovative Activity Change as Industries Mature?', *International Journal of Industrial Organization*, Vol. 19, No. 7, pp. 1141–60.

Nelson, R. R. and Winter, S. G. (1982) *An Evolutionary Theory of Economic Change*, Cambridge, MA: Belknap Press of Harvard University Press.

OECD (1997) *Proposed Guidelines for Collecting and Interpreting Technological Innovation Data: 'The Oslo Manual'*, Paris: OECD and Eurostat.

Palmberg, C. (2006) 'The Sources and Success of Innovations: Determinants of Commercialisation and Break-even Times', *Technovation*, Vol. 26, No. 11, pp. 1253–67.

Palmberg, C., Leppälahti, A., Lemola, T. and Toivanen, H. (1999) 'Towards a Better Understanding of Innovation and Industrial Renewal in Finland: A New Perspective', VTT – Technical Research Centre of Finland, Working paper 41/99, Espoo: VTT.

Pearson, G. J., Pearson, A. W. and Ball, D. F. (1989) 'Innovation in a Mature Industry: A Case Study of Warp Knitting in the UK', *Technovation*, Vol. 9, No. 8, pp. 657–79.

Pesonen, P. (2006) *Innovation Management and its Effect in Forest Industry* (in Finnish), Espoo: VTT Publications 622.

Pesonen, P. and Saarinen, J. (2007) 'The Forest Industry's Innovativeness at Peak Prosperity' (in Finnish), *Paper and Timber*, Vol. 89, No.11, pp. 127–9.

Statistics Finland (2008) StatFin online database service, 2008, accessed in May–September 2008. Available at: http://pxweb2.stat.fi/database/StatFin/databasetree_en.asp

Tushman, M. L. and Anderson, P. (1986) 'Technological Discontinuities and Organizational Environments', *Administrative Science Quarterly*, Vol. 31, No. 3, pp. 439–65.

Utterback, J. M. (1994) *Mastering the Dynamics of Innovation*, Boston, MA: Harvard Business School Press.

Utterback, J. M. and Abernathy, W. J. (1975) 'A Dynamic Model of Process and Product Innovation', *OMEGA*, Vol. 3, No. 6, pp. 639–56.

Windrum, P. (2005) 'Heterogenous Preferences and New Innovation Cycles in Mature Industries: The Amateur Camera Industry 1955—1974', *Industrial and Corporate Change*, Vol. 14, No. 6, pp. 1043–74.

5

Are There Failing Innovations? In Search of Understanding of Failure in Innovation Activities

Nina Rilla and Juha Oksanen

The journey begins

The idea to study innovators' understandings for innovation failure dates back to a research the authors, together with colleagues, accomplished on company innovation processes a few years ago. During intensive periods of data gathering, we carried out tens of interviews with innovators and directors of small and large innovative companies from different sectors across Finland. Most of the interviews were done in pairs and, right after each meeting with innovators and company people, we used to discuss and consider first impressions and insights gained during the interviews.

One of the issues that came to our attention was a seeming discrepancy between stories describing innovation processes from ideation to market and nicely proceeding descriptions given in textbooks on innovation. Innovation process turned out to be a bumpy road, but we had an intuitive feeling that this very lumpiness played critical role in the evolvement of innovation. In some cases there was also an apparent mismatch between tone of innovation stories and prevailing common understanding about the benefits of innovation for company success. Overall, the link between innovation and success turned out to be much more complicated than a naïve innovation researcher anticipated. Evidently, involved actors' understanding about success and failure in innovation were not something which can be measured by using a few 'objective' indicators only. It seemed in many cases difficult, even impossible, to identify and demarcate the area within which success or failure of an innovation process could have been addressed; firms' innovation activities are inseparably linked with firms' business activities and extend in time both towards past and future without succumbing to neat framing of the phenomenon, that is, innovation.

At that moment, we did not have an opportunity to enquire deeper into the ideas born. However, we kept talking about the phenomenon and when an occasion for study innovation failure arose we took it. This chapter is a first account of our efforts to understand what is at stake in innovation failure and how failure relates to innovation, if at all, from the actor's perspective.

The chapter proceeds as follows. The first, introductory section introduces a review of the main studies and concepts of failure in innovation essential for framing the research and provides the aims of the study. Second, the 'Methods' section presents, first, the grounded theory research methodology, which is followed by a detailed description of the analysis process. The third section will introduce our findings, and illustrates how the analyses have evolved from one phase to another, and reflects our findings to the literature. The final section presents concluding comments.

Failure in literature and research objectives

Generally, the literature on innovation activities has concentrated rather on identifying, analyzing and describing success characteristics as opposed to studying innovation failures per se (Henard and Szymanski, 2001; van der Panne *et al.*, 2003). One quite obvious explanation is that succeeding innovations are better known than failing innovations, of which no one really cares to either talk about or document. Moreover, the discussion about success factors has, in fact, been carried out in rather numerical terms, meaning that success has often been assessed with quantitative measures (Henard and Szymanski, 2001).

Some authors approach the success of new products by comparing successful and unsuccessful product development processes (see Rothwell *et al.*, 1974; Maidique and Zirger, 1984; Zirger and Maidique, 1990; Brown and Eisenhardt, 1995; Landry *et al.*, 2008). The others concentrate on evaluating innovation failure by financial measures, such as revenue, sales or pay-back period (Ali *et al.*, 1995; Karlsson and Åhlström, 1999; Palmberg, 2002). On the other hand, some concentrate on assessing factors enhancing success in innovation development – a perspective typical in New Product Development and in management research in general (see Cooper and Kleinschmidt, 1987; Calantone *et al.*, 1995; van der Panne *et al.*, 2003). One of the first comparative studies on the theme of innovation success was the so-called SAPPHO project carried out by Rothwell *et al.* (1974). Their research setting consisted of pairs of successful and unsuccessful industrial innovation cases. In their study the dominant factors discriminating success from failure were: 1) understanding of customer needs; 2) attention to marketing; 3) efficient development process; 4) use of external advice in technology; and 5) experience and seniority of key persons. The innovation success was defined in commercial terms, that is, as obtaining profit.

In contrast to studies examining solely the success or failure of innovation as an outcome, studies that concentrate on examining other aspects of innovation also exist. For instance, the study of Simpson *et al.* (2006) focused on analyzing the negative, along with the positive, consequences of innovation orientation. Their argument is in line with this chapter, indicating that innovation outcome research suffers from positive effect bias. The study suggests that embracing innovation orientation without being aware of negative outcomes, like straying too far from company's core competence, may not lead to desired results.

In addition to innovation literature, organization studies literature and research in the entrepreneurship field provide relevant observations about failure as a phenomenon. According to D'Este *et al.* (2008), literature on organizational change has addressed failure as an enabler for experiential learning. Learning from their own as well other's failing experience has been indicated to have important effects on subsequent activities of companies (Haunschild and Sullivan, 2002; Baum and Dahlin, 2007). On the other hand, Denrell (2003) argues that learning from other organizations is usually based on selection of examples which are skewed towards successful organizations while under-sampling failure. Similarly to organizational change literature, the studies of entrepreneurship provide interesting insights into failing in business activities, which are pertinent to innovation failure. Affinity of entrepreneurship to innovation activity comes through a definition given by Shane (2003) for entrepreneurial activity; innovation as well as entrepreneurship is fundamentally about the discovery and exploitation of opportunities.

Finally, a body of research on project and operations management, focusing, for instance, on research and development (R&D) (Lewis, 2001; Smith-Doerr *et al.*, 2004), software development (Agarwal and Rathod, 2006), implementation of high technology in public safety sector (Dilts and Pence, 2006), project measurement (Ojiako *et al.*, 2008) and company acquisitions (Dalziel, 2008) has addressed explicitly ambiguity of 'success' and 'failure' assessments. A conclusion emerging from these studies indicates that there are not really universal criteria to appraise if a project is a successful or not; rather the perception of success of a project varies among involved actors and stakeholders, depending on a specific context and an individual's organizational position and location in intra- and interorganizational networks.

Despite the literature presented above, we argue that innovation failure in a broader sense is rather unexplored and understanding of failure shallow. Studying failures alongside successes can be seen essential to understanding the dynamics of innovation processes at company level. Naturally, the presented views and factors are valuable for understanding the complex nature of innovation failure but, in our opinion, they are not sufficient. Instead of studying innovation failure by using predefined characteristics and strict theoretical framework, we are interested in examining the question from the actor's perspective, that is, how individual innovators perceive failure in innovation activities, and make sense of innovation failure. Reasons to explore this question include the following. First, common understanding of failure in innovation literature is not explicit. If the topic is in general touched upon, the failure is typically considered as a reverse meaning for success but not as a main research question in itself. Second, the innovators might see innovation failure as a more complex phenomenon than general understanding proposes, but they do not necessarily share prevailing views on innovation failures.

Accordingly, the aim of this chapter is to analyze the understandings for innovation failure that are constructed through innovators' speech in interviews on specific innovation processes. The focus is not only upon a certain time in the innovation process but over a longer period from idea to market and further, enabling us to get a grasp of innovation failure from a longitudinal perspective. Furthermore, we are able to emphasize two important issues in innovation – learning perspective and extended innovation life cycle. By 'extended innovation life cycle' we mean taking into account phases preceding and following a single innovation process. One innovation might be a failure to a certain extent, but prerequisite for the next one to succeed. This study utilizes the grounded theory method, which is seen as an interpretive process, and aims to provide fresh understandings of innovation failure.

Methods

Grounded theory approach

The basis of grounded theory was laid down by Barney Glaser and Anselm Strauss in their book, *The Discovery of Grounded Theory: Strategies for Qualitative Research*, published in 1967. The book presented the first thought-through account for grounded theory research and included the key elements and phases used since then in research – no matter that grounded theory has evolved to various, in their views differing, directions over the time (cf. Mäkelä and Turcan, 2007). A central methodological feature characterizing grounded theory is the view on theory's position in research. Contrary to the quantitative and deductive approaches, theory is not the starting point but the outcome of inductive research process. The analysis process is not directed by theoretical framework but the other way round, and in this way 'theory evolves during actual research and it does this through continuous interplay between analysis and data collection' (Strauss and Corbin, 1988, quoted in Eriksson and Kovalainen, 2008, p. 158). The research process based on grounded theory, despite of its strong inductive basis, involves also characteristics familiar from deductive research – especially when it comes to construct validation and verification procedures (Eisenhardt, 1989; Suddaby, 2006; Eriksson and Kovalainen, 2008). Overall, the research process is characterized by iteration between analysis and data, as well as use of triangulation in order to maximize insights and research quality.

Research drawing on grounded theory is qualitative by nature. Constant comparative method is at the heart of grounded theory approach (Glaser, 2004). The researcher should approach the empirical data openly in analysis phase in order to let the data 'speak' for itself. We agree that existing knowledge cannot be put aside, but it is good to be aware of prior constructs' potential impact for categorization and analysis. Key elements of grounded theory research approach are, first, conceptual categories and their conceptual properties; and second, hypotheses or generalized relations among the categories and their properties (Glaser and Strauss, 1967).

Applicability of resulting theory to practical situations has been an important object of grounded theory approach from the beginning. Glaser and Strauss (1967) emphasize the closeness of constructed theory to process and action – theory should be understandable to the people operating in the area studied. Following the authors (p. 240), we can reformulate the aim of our study as follows: to provide stronger conceptual understanding of failure in innovation context; and to develop concepts and hypotheses to enhance mutual understanding of actors involved in innovation activities (i.e., innovators, management and staff of innovating firms) as well as policy-makers and other stakeholders closely linked to innovation.

Data collection and sampling

Our initial data consists of already-existing interview material gathered between autumn 2005 and autumn 2006. The data was collected by interviewing 80 CEOs or innovators in Finnish innovative companies. Interviews were originally specified on companies' innovation processes, and were recorded and transcribed, which allows the reutilization of data.

In order to keep the amount of data manageable in the start of the analysis and planning for variation, we decided to use some limitation criteria for selecting data. The sampling procedures were directed by two main limitations, of which the first considers the position of interviewee in the firm, and, the second, the size of firm. Since our main aim is to analyze the understandings of innovators, we included only those interviews, in which the interviewee had been the innovator him/herself. This first limitation left us a sample of 61 interviews. As this vast data had to be narrowed down due to research restrictions, we ended up using the firm size measured in number of employees as the second limitation. The innovation activities and procedures are believed to differ between small and large companies, which predicts differing perceptions to failure as well. For this reason, we wanted to ensure the richness of data and make sure that the sample consisted of interviews from several size classes. By using other sampling techniques we might not have succeeded since the majority (63.9 per cent) of our initial sample are micro firms employing less than nine employees.

The sample for this study is composed of thirteen interviews representing views of innovators from micro to multinational companies. The company age varies from three to 31, while the average age is 13.5 years. Another important descriptive characteristic for our sample is the origin of the idea. Only in a few of the cases was the innovation actually based on scientific invention. By 'science-based' we mean ideas originating from scientific research as distinct from non-science-based innovation, which is generally understood to originate from need, either customer or own practices. The background is vital to keep in mind since it is likely to affect challenges innovation activities encounter. The variety of companies' experience in innovation activities and origin of innovative idea increases the richness of the data, which is one of the core issues of grounded theory approach; therefore our size limitation is well justified (see Appendix 1).

The common characteristic to all firms in the sample is that they have been identified as innovative companies. This means they have been responsible for the development and commercialization of a novel product (including tangible and intangible goods and services) – an innovation. Innovation, on the other hand, has been defined as invention that has been commercialized on the market by a company or the equivalent.

Data analysis process

This paragraph will introduce the research process in detail, and aims to bring to the attention of the reader how the analysis has evolved, and how the authors have reached their results. The data analysis process complied with the general procedures of inductive – data-oriented research. The main procedures used comprised of reading, coding, combining and reshaping of transcribed interview data. As the nature of grounded theory suggests, the process has been highly iterative. In different phases of the process we have consciously used the opportunity to analyze material first separately and then compare and combine results in order to increase the accuracy of research. The analysis process of this study is illustrated in Figure 5.1.

After establishing the sample, we started to read through the material. In tandem with reading, we separately coded the material case by case, according to themes we agreed to be relevant in reaching the understanding of failure. These themes were failure and the counterpoint to failure, success. This first phase accounted for 335 different variations of success and failure.

While going through and comparing the separately collected variations and dividing them into categories, we agreed to insert third theme – a challenge – to our analysis. In order to ease the individual case combining process, eleven tentative groups were created. The categorization process continued by comparing the variations. This process included also the reviewing of material once again in order to identify the true meaning of any quotes, so that we were able to classify variations under the correct categories. At this point, the number of categories totalled nine main ones and several subcategories. After preparing the second set of tentative categories, we realized that the level of abstraction was not even nearly reached, which led us to combine and reshape the subcategories. As an outcome of this reshuffling process, 31 new categories, each of which had several properties, were constructed.

In the fifth stage, we started to approach tentative explanatory categories. The categories were analyzed one at a time in order to create relationships between the properties defining the categories. Eventually, this coding stage narrowed the number of categories from 31 to twelve conceptualized categories. Each category contains defining properties, which vary from three up to a maximum of thirteen, depending on category (see Appendix 2). The agreement on the adequacy of the twelve constructed categories initiated the next phase, the process of refocusing our research. Once again, the initial material was read through for new variations to emerge that would justify constructed categories. However,

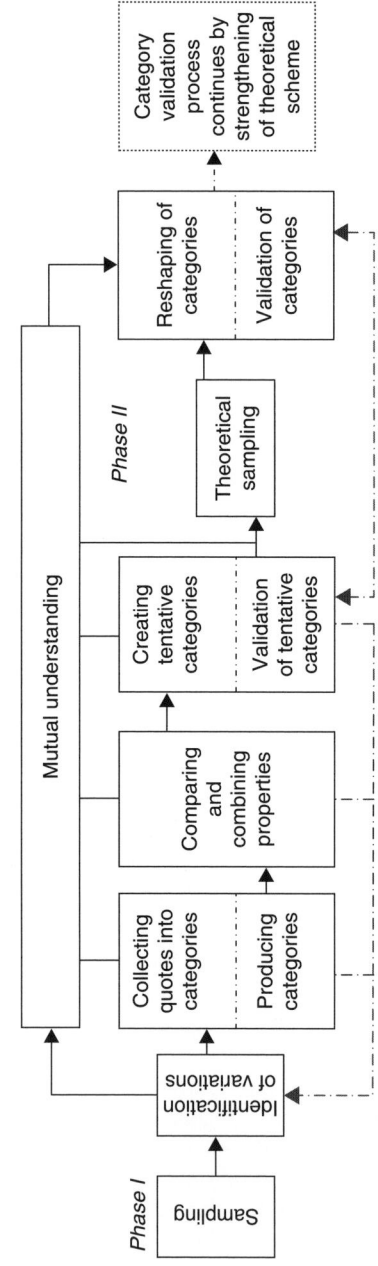

Figure 5.1 The analysis process of the study

in contrast to the first round of reading, this round was directed by the established categories. The validation process is extensive and involves various procedures, of which one is the theoretical sampling to gather new data on a basis of current analysis.

In the second phase of our analysis we aimed at advancing the level of abstraction and creating and strengthening relations between proposed categories. The first step of this analysis round was to select cases for conducting additional interviews, as it was soon noticed that the initial data is not sufficient alone to reach our goal. Given that the criteria for selecting cases in grounded theory is very difficult to determine, we had to rely largely on our own intuition in the judgement in novel aspects and leads which rose from the first analysis. This, on the other hand, is one of the strengths of grounded theory method (Eisenhardt, 1989). Our approach was to collect more data to enable us to elaborate and refine the tentative categories. For that purpose, we selected pertinent cases representing different sides of the phenomenon – success, challenge and failure. The three cases for this second phase were chosen from the original data (i.e., 80 innovators' interview data), and additional interviews were conducted in September 2008. This time the topic of the interviews focused more specifically around the failure theme, giving room for interviewees' own perceptions on failure in innovation activities. At the same time we were able to probe our understanding of innovation failure constructed in the earlier phase. As before, interviews were recorded and transcribed for more detailed analysis, which was carried out in a similar manner as described above in phase I. At the end of this process, known as 'theoretical sampling' in grounded theory parlance, the number of categories had decreased through identifying the most relevant ones, which enabled us to move towards more comprehensive categories.

The analysis strengthened the categories of old variations and categories but, as is the aim of theoretical coding procedure, also introduced completely novel variations for understanding failure. This led us to analyze the current content and relationships of each category, which resulted in producing new categories as well as reshuffling old ones. Consequently, we had fourteen tentative explanatory categories, but at the same time five stronger conceptual categories started to formulate.

Findings and discussion

First steps in understanding innovation failure

As already mentioned above, we had to reformulate our perspective on approaching failure in innovation activities in the first phases of iterative

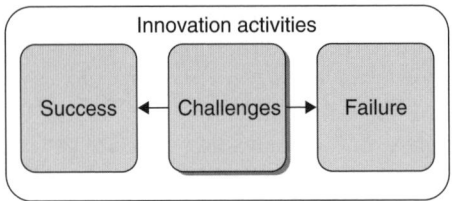

Figure 5.2   Success – Challenges – Failure

process. It became clear that the failure as a phenomenon is not present as such in the material. The concept of failure has to be approached indirectly through challenges and in comparison to success (Figure 5.2). By this we mean that the innovators consider failure in relation to how well they manage to tackle challenges perceived. Second, the innovation activities intertwine seamlessly with entrepreneurial and business activities in our material. Therefore failure in innovation has to be examined in tandem with failure in business activities. These two guidelines drove our analysis process.

The repeated rounds of categorization led to the construction of twelve tentative categories for understanding the dimension through which innovators make sense of success–failure in their innovation activities. Figure 5.3 presents these tentative categories.

At this stage there was no single distinct substantive category that would explain the phenomenon under investigation. However, we identified certain emerging dimensions between the constructed explanatory categories – that is, how the categories relate to each other (Charmaz, 2006). These dimensions shed light on understanding the different aspects of failure phenomenon in innovation and are divided under four themes. The ground-forming categories relate to enabling and motivating ingredients. The second layer consists of categories which emphasize process and action. On the third level are categories that relate to abilities to face the challenges, and, on the other hand, to focus on making critical choices. Finally, on the fourth level, we have placed categories which make demands on success in terms of economic survival as well as fortune. It is worth noting that relations between the categories in Figure 5.3 are described first and foremost on the horizontal line, while the vertical connections were less evident.

The properties relating to the first-level categories underline the significance of motivational and attitudinal aspects, on the one hand, to start and, on the other hand, to continue innovation/business

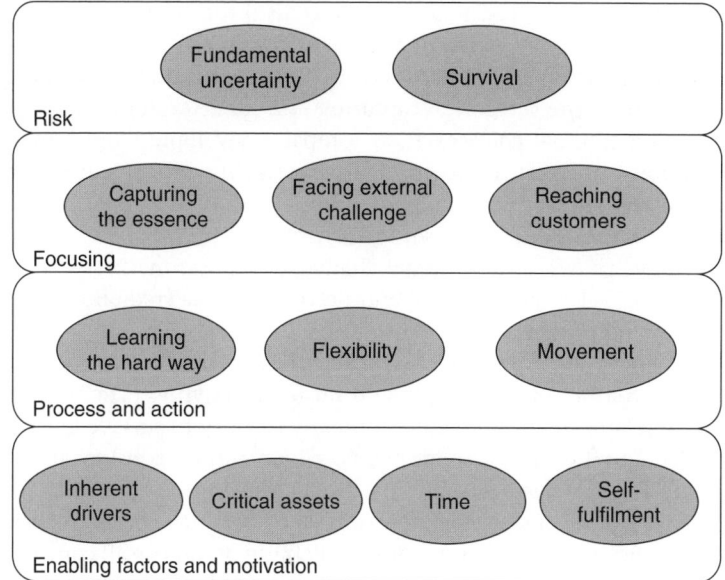

Figure 5.3   Tentative explanatory categories

activities. Without viable crucial assets neither the commencement nor continuation is likely to be possible. The properties in this category include, for instance, personally motivating factors, skills and know-how, contacts and general adequacy and sufficiency of resources. In addition, the timing is an essential element for any of the other aspects to be realized. It is also linked to prospects encouraging future-oriented action.

The role of process and action for success or failure in innovation is characterized by three categories; learning – the – hard way, flexibility and movement. In order to progress there has to be room for making mistakes and to learn. Acknowledgement of the critical importance of learning and tolerance of failure shows clearly when a company is in a pioneer position; there are no well-proven guidelines for how to proceed or which decisions to take. At the same time, one should be flexible and open to both internal and external changes. Moreover, innovation activities require sturdiness and ability to renew.

The potential of a business for innovation depends also on its capability to find the focus, organize operations and make strategic choices about use of resources available while facing external pressure. It is essential to carve out distinct characteristics differentiating the company and its offerings from the competitors, and to distil the essence of activities all the way from idea generation to identification of market segments and reach of customers. Properties related are, at this stage of our analysis, covered by the categories 'capturing essence', 'reaching customers' and 'facing external challenges'. A company developing and bringing innovation to market has to consider strategically critical features of a new (in)tangible product, company product range, as well as organization of operations and processes to stay competitive. At the same time, it may have to overcome external challenges such as market resistance towards a novel product/service and need to achieve credibility among (existing) and potential customers.

Risk dimension includes two categories highlighting the ever-precarious nature of innovation and business activities. The 'survival' category relates to an innovative company's ability to survive and continue its activities in economic terms, which can be considered as an ultimate test for business potential (including innovation) in market economy. Pricing and sales of new developed products, their profitability, cost efficiency of operations and availability of finance all contribute to the success or failure of a business. Yet, notwithstanding the critical importance of economic factors for company survival, it would not be possible without firm's ability to deal with the other aforementioned dimensions of challenge, which a company developing and introducing innovation faces. In other words, survival covers wider aspects than solely financial drivers. More deeply, survival can be comprehended as sustaining the internal drive for self-fulfilment throughout innovation/business activities.

The remaining category highlights the role of 'fundamental uncertainty' always involved in business and development of innovations. As the saying goes, chance favours the prepared mind, but sometimes not even this is enough when new ideas are put on the market for trial. On the other hand, risk is as inherent part of innovation activities as it is of entrepreneurship. Innovators need to be prepared both to take and face risk.

Approaching understanding in innovation failure

The illustration helped us to present the tentative categories, but also provided a mental map for proceeding to the next analysis phase – selective coding. This phase is intended to integrate and refine analysis towards a more focused theoretical scheme.

Given that the initial data was not concerned with innovation failure as such, the additional interviews were essential not only to validate our current interpretations of innovation failure but also to provide novel data. Consequently, the relations between proposed categories started to strengthen, leading to reshuffling of existing and construction of totally new categories (Figure 5.4). All of the categories are in strong relation to time. Time as duration, as well as timing, sets the framework for innovation to occur and, at the same time, fail or succeed. The crucial role of timing in innovation does not include here any normative assumptions, such as first mover's competitive advantage in market: in fact, depending on a specific situation and involved actors, pioneer position may well be a disadvantage.

As mentioned above, after the interviews it became clear that as innovation is realized through business, innovation in innovators' minds is impossible to separate from business activities. Therefore, orderly

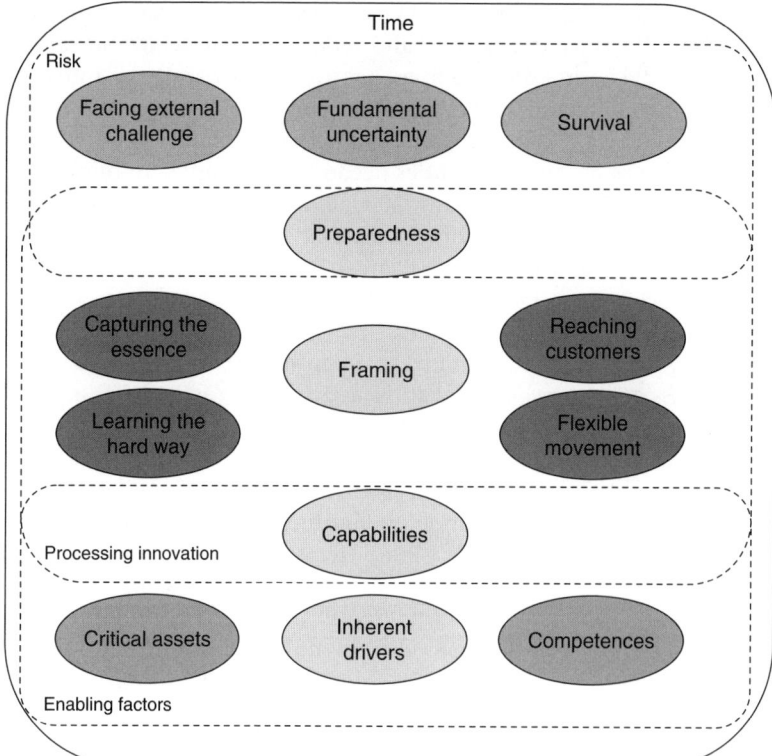

Figure 5.4 Explanatory categories in innovation failure

management can to some extent prevent innovation failure, as wide literature on success factors also implies (van der Panne *et al.*, 2003).

Given that innovation is intertwined with business activities, part of innovation failure can be understood as factors which innovators are able to impact with their actions or with proper preparedness. For instance, the success factors, like understanding of customers and experience of key persons, identified in the study by Rothwell *et al.* (1974) were also emphasized in our material. However, the counterpart of orderliness is non-orderliness, space for creativity – a central, even necessary, dimension of innovation and innovation activities. In that sense, innovation process can be seen as a balancing act or dialogue between free idea creation and processes characterized by orderliness. The former supports creation, while the latter supports identification of innovative ideas and enables them to be taken further.

Competences and people are significant in innovation but most of all capabilities matter, such as ability to learn from setbacks and tolerate failure. Critically assessing and reflecting own behaviour is one of the key elements of capabilities. On the other hand, the inherent drives go hand in hand with capabilities. These findings are in line with the vast literature of dynamic capabilities, that emphasize the ability to combine the assets and capabilities needed, for example, in innovation activities (Eisenhardt and Martin, 2000) and in organizational learning (Lane and Lubatkin, 1998). Without motivation, commitment and commonly shared goals innovation is likely to fail. Not only maintaining but also enhancing the inherent drivers throughout innovation activities is significant in innovation's viability.

Finding the focus of innovation can definitely be considered as a learning process. The core of innovation activities consists of finding the essence of both innovation and capabilities. Balancing between the challenges of creating demand for innovation and developing innovations to supply demand is one of the key aspects in innovation activities. Framing in every aspect of innovation and business activities is therefore an important factor in innovation failure. The essence of framing is not only seen in flexibility to adjust one's capabilities but also as preparedness to face challenges and uncertainty and to tap emerging opportunities. Most of all framing is understood to be linked to setting the context for innovation – finding the way of doing the right things. In fact, experience gained from failure can turn out to be more significant in finding the core competences than learning from success (D'Este *et al.*, 2008).

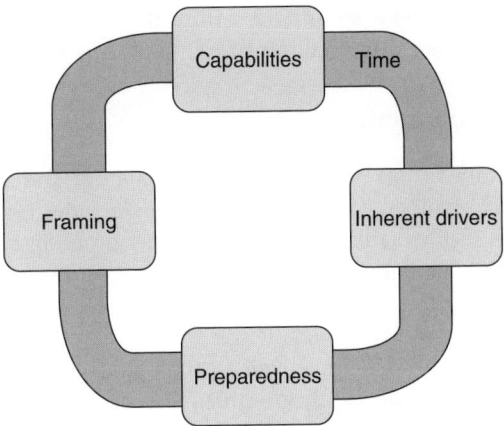

Figure 5.5 Substantial categories in innovation failure

Being prepared includes flexibility; it is not just about avoiding failures but more about preparedness to handle failure mentally, as well as in business. Failure in innovation cannot be excluded either; it is essential since without experience of failing you are likely to fail. Therefore failure is not understood as a concrete incident but as a challenge. A recent study on entrepreneurs' attitudes towards failure (Politis and Gabrielsson, 2007) addresses the importance of a positive attitude towards failure, and, most of all, experiential learning in entrepreneurial process, which were also issues emerging from our analysis.

Different types of risk and uncertainty in innovation activities prohibit companies from innovating and lead to high failure rates (van der Panne *et al.*, 2003). The innovation process is complex and its viability is dependent on various factors (van der Panne *et al.*, 2003). Often viability is not even contingent on concrete factors, but actions or actors, or simply serendipity. As several studies reveal (the majority of which have concentrated, however, on analyzing the management of success factors), innovation success or failure may happen at firm, project, product or market level. Entrepreneurship processes, as well as innovation activities, are ultimately about uncertainty (Shane, 2003). The unpredictability of future challenges and opportunities makes innovators alert to changes, in which adequate amounts of flexibility and adaptability become crucial factors. Flexibility and preparedness are, of course, needed in organizational, entrepreneurial and innovation levels. Figure 5.5 highlights the five substantial concepts summarized from the above discussion: time, capabilities, inherent drives, framing and preparedness.

Concluding comments

Engaging in innovation means taking the challenge and accepting the uncertainty it involves; taking the challenge includes possibilities for succeeding as well as failing. Equally, the phenomenon of innovation failure has been demanding for researchers to get a grasp of. As our analysis shows, the failure concept in innovators' understanding is deeply intertwined with success and challenge.

Taking into account the novel and abstract character of our topic, we chose to tackle the issue with grounded theory approach. Our main aim was to analyze the failure in a practical context, that is, from the actor's perspective, and to bring out the interpretative flexibility and the diversity around the concept of innovation failure.

Based on close reading and analysis of interview data, we introduced five substantial categories, which represent the dimensions for how innovators understand and make sense of failure in innovation activities. On the one hand, these categories can be understood as challenges, which were constructed from the initial data, that is, from the innovator's sense-making of innovation process evolvement. On the other hand, these challenges have evolved into extensive concepts, providing more abstract understanding for innovation failure. We are not proposing the aforementioned categories to be the critical means to success, but they are vital aspects of business activities, as perceived by innovators themselves.

The research setting is relatively novel in innovation management studies, and therefore brings interesting and beneficial results to evaluating innovation activities, from a management as well as from a policy perspective. The results contribute the understanding for the innovation failure phenomenon – as new insights emerged but also old ones strengthened. Above all, this study provides a starting point for future thorough examination on the complex theme of innovation failure. Hopefully, the results will assist in bringing the views of various actors, for example, innovators, funding agencies and investors, closer to each other by increasing the awareness of complexity around innovation failure. This view stresses the need to take into account individual perceptions, as well as individual objectives and targets. Failure is always largely a context-dependent, not black and white phenomenon, as also indicated by the study of Smith-Doerr *et al.* (2004).

So, are there failing innovations? At the moment, as our research journey reaches its end, and after lengthy and profound discussions, we are convinced that failure exists and is even an essential element in innovation activities. Failure has, however, both negative and positive aspects – failure at a certain moment may precede later success. Successful innovators are able to turn setbacks into beneficial experiences for progress. As one innovator said, 'bypaths are just compulsory in order to find the core [of innovation]'.

Limitations

The size of our sample, including thirteen initial interviews with innovators and company CEOs and three additional focused interviews in the second phase, is still rather restricted for a grounded theory study dealing with such an ambiguous issue as failure in innovation (e.g., Charmaz, 2006). We strongly feel that additional interviews on the theme of innovation failure and, in particular, inclusion of other stakeholders besides innovators would further strengthen our analysis. Furthermore, additional data would allow us to ensure whether the substantial categories arrived at in the analysis are conceptually dense enough to account for innovators' understanding of innovation failure in diverging circumstances. Therefore, the results of the analysis are first and foremost indicative rather than conclusive.

In a methodological sense, the use of existing interview data collected for other purposes is a possible weakness in our effort to follow grounded theory approach when studying failure in innovation context. According to the literature, overlap between the data collection and analysis phases is one of the key strengths in grounded theory research and in qualitative research in general. The iterative and concurrent research process provides a researcher with an opportunity to react swiftly on issues and themes emerging in analysis, adjust the work plan accordingly and collect additional data in order to achieve deeper understanding of the phenomenon studied (Eisenhardt, 1989; Suddaby, 2006).

Appendix 1: Descriptive information of sample

	Position of interviewee/s	Year of establishment	Size (no. of employees)	Origin of innovation	Type of innovation
Phase 1					
F1	(1) innovator/marketing and sales manager (2) innovator/technical advisor	1987	100–249	demand	product
F2	(1) innovator/CEO/founder (2) innovator/R&D manager	1977	10–19	science	product
F3	innovator/R&D manager	2002	10–19	science	product
F4	innovator/CEO	1984	1–4	demand	product/service
F5	innovator/CEO	1998	1–4	demand	product
F6	innovator/CEO	2002	1–4	demand	product
F7	innovator/CEO	1997	20–49	science	process/product
F8	innovator/ex-CEO	1999	250–499	demand	product
F9	innovator/category manager/ex-entrepreneur	1974	50–99	demand	product
F10	innovator/ex-CEO/ founder	1995	5–9	demand	product
F11	(1) innovator/CEO/ founder (2) innovator	1995	5–9	science	process
F12	innovator/CEO/ founder	1990	5–9	demand	process/product
F13	innovator (retired)	1989	999+	demand	product
Phase 2					
F14	(1) innovator/CEO (2) innovator/sales director	1997	20–49	demand	product
F15	innovator/CEO	1994	10–19	science	product
F16	innovator/CEO	2002	EXIT (2008)	demand	product

Appendix 2: The first exercise for grouping of categories

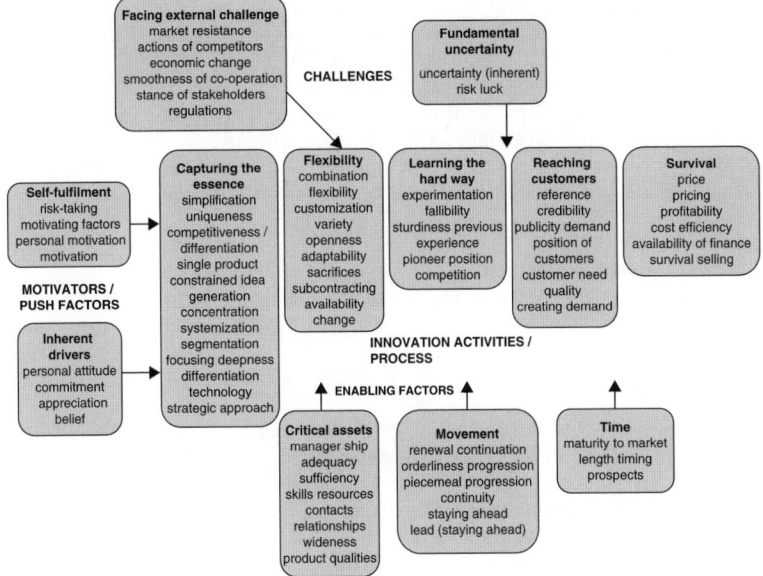

References

Agarwal, Nitin and Rathod, Urvashi (2006) 'Defining "Success" for Software Projects: An Exploratory Revelation', *International Journal of Project Management*, Vol. 24, pp. 358–70.

Ali, Abdul, Krapfel, Robert Jr and LaBahn, Douglas (1995) 'Product Innovativeness and Entry Strategy: Impact on Cycle Time and Break-even Time', *Journal of Product Innovation Management*, Vol. 12, No. 1, pp. 54–70.

Baum, Joel A. C. and Dahlin, Kristina B. (2007) 'Aspiration Performance and Railroads' Patterns of Learning from Train Wrecks and Crashes', *Organization Science*, Vol. 18, Nos. 3, pp. 368–87.

Brown, Shona L. and Eisenhardt, Kathleen M. (1995) 'Product Development: Past Research, Present Findings, and Future Directions', *Academy of Management Review*, Vol. 20, No. 2, pp. 343–78.

Calantone, Roger J., Vickery, Shawnee K. and Dröge, Cornelia (1995) 'Business Performance and Strategic New Product Development Activities: An Empirical Investigation', *Journal of Product Innovation Management*, Vol. 12, No. 3, pp. 214–23.

Charmaz, Kathy (2006) *Constructing Grounded Theory. A Practical Guide Through Qualitative Analysis*, London: Sage.

Cooper, Robert and Kleinschmidt, Elko (1987) 'New Products: What Separates Winners from Losers?', *Journal of Product Innovation Management*, Vol. 4, No. 3, pp. 169–84.

Dalziel, Margaret (2008) 'The Seller's Perspective on Acquisition Success: Empirical Evidence from the Communications Equipment Industry', *Journal of Engineering and Technology Management*, Vol. 25, pp. 168–83.

Denrell, Jerkel (2003) 'Vicarious Learning, Undersampling of Failure, and the Myths of Management', *Organization Science*, Vol. 14, No. 3, pp. 227–43.

D'Este, Pablo, Iammarino, Simona, Savona, Maria and von Tunzelmann, Nick (2008) 'What hampers innovation? Evidence from the UK CIS4', SEWPS, SPRU Electronic Working Paper Series, Paper No. 168.

Dilts, David M. and Pence, Ken R. (2006) 'Impact of Role in the Decision to Fail: An Exploratory Study of Terminated Projects', *Journal of Operations Management*, Vol. 24, pp. 378–96.

Eisenhardt, Kathleen M. (1989) 'Building Theories from Case Study Research', *Academy of Management Review*, Vol. 14, No. 4, pp. 532–50.

Eisenhardt, Kathleen M. and Martin, Jeffrey A. (2000) 'Dynamic Capabilities: What are They?', *Strategic Management Journal*, Vol. 21, Nos. 10/11, pp. 1105–21.

Eriksson, Päivi and Kovalainen, Anne (2008) *Qualitative Methods in Business Research*, Thousand Oaks, CA: Sage.

Glaser, Barney G. (with assistance of Judith Holton) (2004) 'Remodelling Grounded Theory', *Forum: Qualitative Social Research*, Vol. 5, No. 2, Art 4, May.

Glaser, Barney G. and Strauss, Anselm (1967) *The Discovery of Grounded Theory: Strategies for Qualitative Research*, New York: Aldine.

Haunschild, Pamela R. and Sullivan, Bilian Ni (2002) 'Learning from Complexity: Effects of Prior Accidents and Incidents on Airlines' Learning', *Administrative Science Quarterly*, Vol. 47, No. 4, pp. 609–43.

Henard, David H. and Szymanski, David M. (2001) 'Why Some New Products are More Successful than Others', *Journal of Marketing Research*, Vol. 38, No. 3, pp. 362–75.

Karlsson, Christer and Åhlström, Pär (1999) 'Technological Level and Product Development Cycle Time', *Journal of Product Innovation Management*, Vol. 16, No. 4, pp. 352–63.

Landry, Réjean, Amara, Nabil and Becheikh, Nizar (2008) 'Exploring Innovation Failures in Manufacturing Industries', paper presented at the 25th DRUID Celebration Conference, 17–20 June, Copenhagen.

Lane, Peter and Lubatkin, Michael (1998) 'Relative Absorptive Capacity and Interorganizational Learning', *Strategic Management Journal*, Vol. 19, No. 5, pp. 461–77.

Lewis, Michael A. (2001) 'Success, Failure and Organisational Competence: A Case Study of the New Product Development Process', *Journal of Engineering and Technology Management*, Vol. 18, pp. 185–206.

Maidique, Modesto and Zirger, Billie Jo (1984) 'A Study of Success and Failure in Product Innovation: The Case of the US Electronics Industry', *IEEE Transactions on Engineering Management*, Vol. 31, No. 4, pp. 192–203.

Mäkelä, Markus and Turcan, Romeo (2007) 'Building Grounded Theory in Entrepreneurship Research', in Helle Neergaard and John Ulhøi (eds), *Handbook of Qualitative Research Methods in Entrepreneurship*, Cheltenham: Edward Elgar.

OECD (2005) *Oslo Manual: Guidelines for Collecting and Interpreting Innovation Data* (3rd edn), Paris: OECD and Eurostat.

Ojiako, Udechukwu, Johansen, Eric and Greenwood, David (2008) 'A Qualitative Re-construction of Project Measurement Criteria', *Industrial Management & Data Systems*, Vol. 108, No. 3, pp. 405–17.

Palmberg, Christopher (2002) *Successful Innovation: The Determinants of Commercialisation and Break-even Times of Innovations*, VTT Publications: 486, Espoo: VTT Technology Studies.

Politis, Diamanto and Gabrielsson, Jonas (2007) 'Entrepreneurs' Attitudes Towards Failure: An Experiential Learning Approach', Babson College Entrepreneurship Research Conference (BCERC). Electronic copy available at: http://ssrn.com/abstract=1064982

Rothwell, R., Freeman, C., Horsley, A., Jervis, V. T. P., Robertson, A. B. and Townsend, J. (1974) 'SAPPHO Updated-Project SAPPHO Phase II', *Research Policy*, Vol. 3, pp. 258–91.

Shane, Scott (2003) *A General Theory of Entrepreneurship: The Individual–Opportunity Nexus*, Cheltenham: Edward Elgar.

Simpson, Penny, Siguaw, Judy and Enz, Cathy (2006) 'Innovation Orientation Outcomes: The Good and the Bad', *Journal of Business Research*, Vol. 59, pp. 1133–41.

Smith-Doerr, Laurel, Manev, Ivan M. and Rizova, Polly (2004) 'The Meaning of Success: Network Position and the Social Construction of Project Outcomes in an R&D Lab', *Journal of Engineering and Technology Management*, Vol. 21, No.1, pp. 51–81.

Suddaby, Roy (2006) 'From the Editors: What Grounded Theory is Not', *Academy of Management Journal*, Vol. 49, No. 4, pp. 633–42.

van der Panne, Gerben, van Beers, Cees and Kleinknecht, Alfred (2003) 'Success and Failure of Innovation: A Literature Review', *International Journal of Innovation Management*, Vol. 7, No. 3, pp. 309–38.

Wilson Scott, Karen (2004) 'Relating Categories in Grounded Theory Analysis: Using a Conditional Relationship Guide and Reflective Coding Matrix', *Qualitative Report*, Vol.9, No. 1, pp. 113–26.

Zirger, Billie Jo and Maidique, Modesto A. (1990) 'A Model of New Product Development: An Empirical Test', *Management Science*, Vol. 36, No. 7, pp. 867–83.

Part III
Sectoral Aspects of Innovation

6
Innovation and Dynamic Strategy: Planning and Implementing Continuous Renewal

Pekka Pesonen

Introduction

While pursuing innovations and higher profits, firms struggle with the growing challenge of planning successful innovation activity for today, let alone for tomorrow. In order to determine the innovative actions of tomorrow, companies try to monitor the external, as well as the internal changes of their business and predict them. Individual innovations, in turn, are answers for the monitoring and prescribed in the innovation activity plan, that is, innovation strategy. Since innovation is no option nowadays, but more of an imperative, it is necessary to comprehend the relation that innovation and technological development have to the changes in firm's internal and external circumstances. By understanding this linkage, organizations can be more capable in steering their *innovation strategy* and creating successful innovations.

Previous studies, however, on innovation and strategy-making have not been explicit in linking the two. Studies in strategic management of a firm from the innovation perspective are very scarce. Individual innovations have been discussed in a more abstract level and in isolation from the management of a firm, rather than dealing with the nature and the characteristics of single innovations in relation to the management of a firm. Thus, the understanding of single innovations being the results of innovation strategy has not captured needed attention, resulting in a knowledge gap between the plan and the outcome. Understanding the present external situation and the internal characteristics, as well as the dynamics of these factors, firms can better continuously reshape the strategy along with technological development. As put by Mintzberg (1994), strategy-making process is a fundamentally dynamic one, corresponding to the dynamic conditions that drive it.

In this chapter I argue that, while innovation is a central ingredient for firm success, in order to guarantee triumph in the longer run innovation needs to be understood in a wider organizational context; in concert with the firm's position in the competitive setting and with the anticipated changes in it, forging firm's innovation strategy and thus the nature of individual innovations the firm aims to generate. Furthermore, I argue that, as innovation, innovation strategy and strategic position (i.e., the internal and external situation of a firm) are interlinked, there is a necessity for dynamic, continuous actions in formulating innovation strategy.

Markides (1999) described the elements of a dynamic business strategy in terms of changing strategic positions of the firm, while Chiesa and Manzini (1998) spoke of competence-based dynamic technology strategy. Both of these studies took an action point of view where the results showed how to change a strategy from one to another and with what type of actions. The present chapter has a different approach. It tries to link the technological change in terms of *innovations* with the strategic position of an organization and discover how their interaction takes place through *dynamic strategy-making*, rather than specifying different strategies or steps in creating them.

I apply SWOT (strengths/weaknesses/opportunities/threats) as a tool for describing the strategic position and its dynamics in relation to innovation. The purpose of the chapter is not to generate a new model of SWOT or to give new acronym for the old one, but simply to create understanding on the dynamics of the strategic position of an innovating firm in an industry. This is a crucial issue for managers as they try to clarify the upcoming changes in strategic position that they need to pursue, and the ones that they should not aim at. The chapter addresses how innovation interacts with firm's strategic position in the business environment. This is studied in the context of formulating innovation strategy, which is understood as defining the innovations to be developed by a firm. As an example, the case of Finnish forest industry is used.

Theoretical background

Strategic management

A strategy of a firm is generally understood as consisting two basic components: the goals or visions a firm pursues and the means it uses to try to attain them. The process of strategy formation, by which a firm develops its strategy, usually includes some similar elements. The most frequently appearing idea, as argued and criticized by Mintzberg (1990), in theoretical literature of strategy-making is the fit between external factors and organizational factors. In practice, this fit means that, after setting the objective in strategy, firms most often asses the external and internal conditions of the organization. Although they are described a bit differently in detail,[1] the fundamentals of these internal and external assessments in strategy formation are the same; evaluating the strengths and weaknesses of the organization on the one hand, and evaluating the opportunities and the threats of the environment on the other.

A similar duality of assessed issues in the beginning of strategy creation was also what Porter (1980) displayed. He noted that competitive strategy is formulated in the context of two aspects. First, it includes the strengths and weaknesses combined with personal values which determine the internal limits to the strategy a company can successfully adopt. Second, industry opportunities and threats combined with broader environment define the risks and potential rewards of the competitive environment, and wider external limits to the strategy.

But the big question that arises from this line of thought is how to do this. What is needed in evaluating the internal and the external situation, and is it so easy? In order to analyze the most likely opportunities, threats and future position of a firm or an industry in general, one has to understand the evolution of a business environment. As put by Ansoff (1984), in strategic planning, the future is not necessarily considered better than the present and thus deemed as not possible to forecast by extrapolating. Because firms cannot predict the future, they have to analyze it by estimating the possibilities, trends, threats and single events leading to change in order to anticipate what the future might be. But understanding the nature of technological change – which is deemed as one important issue to estimate – and single events (i.e., innovations) driving it, is problematic in many ways. One major problem is that managers do not necessarily have the capabilities to do it and thus cannot interlink technological change into strategic planning and strategy formulation. This is due to the fact that some key management decision-makers might have inadequate background (e.g., education) and ability to make judgements and forecasts in the area of technology (Fusfeld, 1978). This in turn leads to a situation where the importance of technological issues in strategy is downgraded.

Shaping strategy is not an everyday task in companies, or even a cyclical one. In most cases, strategy gets changed because something fundamental has changed in the environment, on a one-time basis (Mintzberg, 1994). This happens especially with the occurrence of external discontinuities rather than gradual changes. Thus, in practice, organizations do not necessarily reshape their strategy cyclically, or even regularly, but more in the sense of answering to a changed external situation. Hence, it seems that the strategy-making is not in many cases a dynamic, not to mention a proactive process, but a rather reactive one. This poses a second question relating to the formulating of a strategy. When and how often should firms shape their strategy? If it should be done more dynamically, is there a need for a long-term strategy at all? Regarding the issue of the time span of the strategy, there are different arguments. Minzberg (1994) argues that planning can be long term only when the environment cooperates,[2] because strategic planning is always extrapolating the known trends of the present. Hamel and Prahalad (1994), on the other hand, say that there are long-term and short-term plans made separately, which in practice should be considered as a whole and put together, because the 'long term starts now'. Long-term strategies are hard, or even impossible, to implement as such in the far future because the environment is more complicated to predict. As it is evident that environmental factors evolve over time, their influence in creating a dynamic strategy is deemed very significant (see Porter, 1991). In that sense, a long-term strategy cannot be static because of the risk of locking a company into past experiences and established practices in the evolving environment (Reinmoeller, 2002). Relying too much on long-term plans can put a company on hold in technological development, but having some idea about the objectives and ways to reach them in the longer term is necessary in order to create more radical innovation (see Hamel and Prahalad, 1994). It is also clear that not every change in the environment happens in the short term, especially in more static and mature industries. Hence, plans for longer time spans are needed as well, but, as mentioned before, the strategy formation should be a dynamic process reacting or even being proactive to the external, as well as internal, situation. To be continuously competitive, the long-term strategy needs to be modified as well, as there is a need for new short-term plans.

In addition to the interaction between strategy and the evolving business environment (external situation) there is interaction between strategy and internal factors. These internal factors can be, for instance, resources, skills, competences or capabilities. Regardless of the form of the internal factor, technology is probably the most fundamental asset behind all of the mentioned characteristics. The interaction between strategy and technology can have basically three different perspectives according to Itami and Numagami (1992). The most obvious one is that current strategy is formulated in the sense of that it exploits the technology at hand in the best possible way and within its limits. This is because, in many cases, the technology of the firm is its main resource and thus the main goal of the strategy (especially in the short term) is to benefit from it. The present industry's (or competitors') technology acts also as a threat, thus affecting the strategy-making (Porter, 1991). The second perspective of strategy–technology interaction is that strategy cultivates technology, that is, present strategy steers the development of the technology of the firm. Finally, technology drives the cognition of the strategy, meaning that the current technology affects the future strategy of the firm. Too strict commitment to the old technology can hinder innovation, but it can also create it. This can be the case, for example, when many people share a deep understanding of a certain technological area driving idea generation, and also making it easier for the organization to adapt new strategic directions.

Innovation strategy

The aspect of technology affecting to the strategy is also dealt with in the theories of technological trajectory and path dependency (see Dosi, 1982; Teece *et al.*, 1997; Tidd *et al.*, 2001). These aspects note that a firm's innovation strategy is constrained by the specific opportunities open to it in the future, while the present position of a firm is shaped by the path it has travelled. Technological knowledge of a firm depends on its developed products, and learning and accumulating new knowledge is related closely to previous activities and routines (see Nelson and Winter, 1982). Another constraint is a firm's competences, in other words, what it is capable of learning and exploiting. Learning tends to be incremental, since major steps increase uncertainty and reduce the capacity to learn, leading a firm to follow a more evolutionary trajectory. As argued by Cohen and Levinthal (1990), the ability to recognize and assimilate new knowledge is a function of a prior-related knowledge, which they labelled as 'absorptive capacity'. Thus firm's new technology is often built on existing knowledge, placing a firm on a certain technological trajectory, which Dosi (1984, p. 15) defines as 'the pattern of "normal" problem solving activity (i.e. of "progress") on the grounds of a technological paradigm'.

Innovation strategy guides the innovation activity of a firm by specifying *what* kind of innovations, in terms of technological, non-technological and organizational changes, are to be developed, *how* the innovation process is carried out and to *whom* the innovations are targeted. Outcomes of the strategy are changes and developments in firm's action and offerings. Basically, we are talking of innovation, which refers to a change in the products or services the company offers to markets or a change in the way these offerings created and delivered (Tidd *et al.*, 2001). Moreover, innovation can take different forms. Traditionally the focus has been on product and process innovations, but widening definition has taken into account, for instance, services, organizational and market innovations as well (see OECD, 2005). Despite its many forms, innovations often involve technology in their architecture, or they affect to the way new technology is created. Thus, purely technological innovations not only advance technological development, but also non-technological ones. For instance, a service innovation usually has a technological aspect in the process of producing or distributing the innovation (e.g., internet bank services), or it can support the creation of new technological innovation (e.g., R&D consultancy).

Innovation changes the internal competences and characteristics of a firm because, throughout the innovation process, the firm develops new skills and knowledge related to the innovation and its production. It also shifts the position of the firm in relation to its environment, for example, industry's technological trajectory, as the firm develops its internal technology. The phenomenon occurs also the other way around when the technology of the industry or sector is developing and changing the firm's environment and thus the lens through which the company perceives itself. This means that the internal competences and firm characteristics, and the external environment are evolving somewhat in interaction. This is also noted by Clarke and Thomas (1990) in the case of external possibilities. They say that the external technological change of industry produces certain technological opportunities around which firms and managers make choices. At the same time firms' technological competences are reshaping those opportunities and the decision-making process by which those opportunities are evaluated and exploited.

The situation is not easier when it comes to evaluating one's internal strengths and weaknesses, nor is it easy to combine these two perspectives in strategy formation; the external situation and its probable changes, and internal competences and weaknesses. Managers might not have a perfect or even necessary knowledge of the internals of the firm to formulate a successful strategy. This is highlighted in innovative, fast-changing technological fields where there is no experience about the competences of today, let alone of tomorrow (see Tidd *et al.*, 2001, p. 71, for a case example).

The discussed dynamic interaction between technological development, internal firm characteristics and external environment, and strategy is illustrated in Figure 6.1. The linkage of strategic position and innovation takes place through innovation strategy, in which a

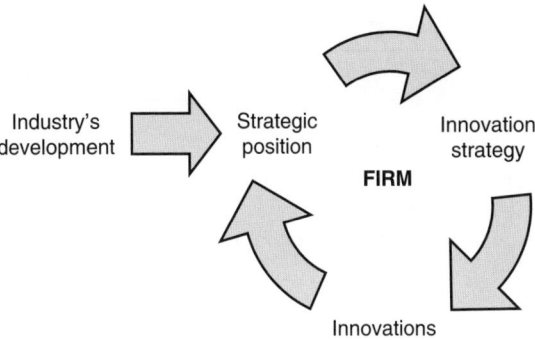

Industry's development	→	Strategic position	Innovation strategy

FIRM

Innovations

Figure 6.1   Dynamics of innovations, strategic position and innovation strategy

firm defines its desirable innovation activities. Innovation activity aims at commercializing new knowledge innovations. New innovations, in turn, together with the industry's technological development, shape the strategic position of a firm. As various internal and external factors change, their evaluation does not remain valid over time, but it needs to be done more dynamically. This creates an undoubted need for reshaping the innovation strategy dynamically as well. The dynamic pattern illustrated provides a frame for the empirical part of the chapter.

The pattern of innovation and strategic position

The case industry

The studied unit in this case is the Finnish forest industry. Its competitive environment is understood as consisting of both foreign forest industry companies and companies in other industries competing in the same markets/products. It is acknowledged that there are many product categories – even very different from each other – in the forest industry, and they might have differences in competing environment and business logics. However, this study does not focus precisely on one product, but on forest industry firms and products in general. The industry's various product areas and markets have several similarities in characteristics that will be discussed in forming the strategic position of firms. Examples of innovations will be used from different areas of the industry, just to describe a phenomenon or to illustrate a point, not to give universal rules.

In order to study if there has been a change in the external environment and the internal position of the firms in the industry at hand, two

Table 6.1 Key figures of the Finnish forest industry[3]

	1991	2006
Production (mill. m^3)		
Paper	8,8	14,2
Pulp	8,1	15,9
Mechanical forest industry	6,9	14
Share of exports in production (per cent)		
Paper	86	90,1
Pulp	17	32,7
Mechanical forest industry	70,8	66,4
Share of employment (per cent)		
Total forest industry	3,4*	2,6
Share of GDP (value added) (per cent)		
Total forest industry	4,4*	3,7
Share of exports (per cent)		
Total forest industry	38	20,1

* 1990.

different SWOTs are illustrated. First, SWOT describes the situation in 1991 and the second in 2006 (see Appendix). In addition, some key values describing the industry and giving more comprehensive background information of its changes are presented. Second, the empirical part studies how the SWOT and companies' innovation strategy has interacted and how have firms reacted (or been proactive) to the changing situation.

The production of forest industry goods has increased radically in the last fifteen years (Table 6.1). The mechanical forest industry has doubled its production and pulp and paper is not far from this rate of increase either. The growth of these figures can be explained by big investments in the 1990s to new machines and factories, which led to boosted domestic production capacity. Capacity growth was also due to rapidly developing processes through innovations increasing automation, speed and reliability of production lines in factories.

The export shares of different products show the increased importance of exporting the chemical forest industry's products (pulp and paper), while at the same time sawn wood and other wood-based products are relatively more domestically consumed. Because the domestic markets in Finland are rather small, the exported paper products have become increasingly important for firms in global competition. At the same time the importance of the forest industry for the Finnish economy has decreased, indicated by the share of GDP and the share of total exports.

Evolution of SWOT in the Finnish forest industry

As the two SWOT tables of 1991 and 2006 (see Appendix) show, the competitive situation of the industry was actually quite different fifteen years ago compared to the present day. One major change in the business environment of the FFI (Finnish Forest Industry) has been the transferring locus of growing markets. The Finnish forest companies concentrated on European markets in 1991 and had some 84 per cent of the exports directed there, having as much as 20 per cent share of the market size. In 2008 the growth in closely located European markets had decreased and the supply of paper-making in Europe had exceeded the demand, creating ever more tough competition.

In 1991 the FFI was technologically the most advanced in the world, having high production capacity as a competitive strength. Now, this competitive advantage has diminished as foreign companies have at least to some extent closed the technological gap. Still the FFI has some advantage in quality, efficiency and technology in general, based on an advanced national innovation system, good education system, high R&D investments in the forest sector, as well as ambient forest cluster actors with state-of-the-art technology enhancing the development of FFI. In 2006, different actors in the value chain and related industries of the FFI worked closely together, having a national innovation strategy in order to create new products and technologies. The relative technological advance of the FFI has, however, decreased in fifteen years, altering the focus of innovation strategy also towards new products in addition to process developments.

An interesting issue in the FFI's SWOT is the role of raw material, and especially domestic wood. In 1991 it was considered basically as a strength; yearly felling rate was only 65 per cent of total growth and the FFI was mainly self-sufficient regarding domestic, logistically easy-to-get and high-quality wood. In 2008, raw material and issues related to it can be found in every corner of SWOT. The felling rate was practically as high as possible (85 per cent) and lack of domestic wood is raising its price, as well as forcing companies to import more and more timber, which is becoming more expensive because of tariffs in Russia. Wood as a raw material has maintained its strength in the light of its renewability and environmental friendliness. It gives also new opportunities for companies to innovate, for instance using logging waste as a source of biofuel or recycling wood-based products to energy and raw material (e.g., recycled paper to produce new paper). As the importance of recycling, other environmental issues have also become more important factors in the competitive environment of the FFI and changing the SWOT.

Analysis of the competitive situation of the FFI over a time period of fifteen years shows that there have been major changes in both internal characteristics and external environment of the industry. As described above, a number of individual changes are consequences of technological development and innovation (e.g., advanced process technology). Many of the changes in competitive factors in SWOT and their significance has also influenced the innovation activity of the firms, that is, they can be seen as a cause or drivers for innovation (e.g., increased collaboration in forest cluster). Thus, it seems that innovation and the changing SWOT have a close linkage, and the interaction between these elements occurs both ways. In the next chapter, the interaction of innovation and SWOT is more thoroughly studied.

Innovation through the changing lens of strategic position

One might argue that, in practice, SWOT as a strategy tool is outdated and out of use. However, for instance, in Finland SWOT is the most widely used tool among the executives in the 500 biggest companies to support the decision-making and planning activities (Stenfors and Tanner, 2007). It is also regarded as one of the cornerstone ideas in strategic management due to its impact, utility and longevity (Allio, 2006). Its application is in the identification and analysis of internal and external environments in order to support the strategic decision-making. It is a clear way to generate understanding, in an easy and applicable[4] manner, on the position of the studied unit, and thus a good basis for successful strategy formation.

Figure 6.2 illustrates the dynamics of SWOT related to innovation. The example factors in each of the four boxes of the analysis are just samples of longer lists of identified factors in the SWOT of Finnish forest industry, which is illustrated in the Appendix. Presented factors are put in Figure 6.2 to help in explaining and understanding the supposed pattern of SWOT. The unbroken lines in the figure represent the goals of innovation strategy, that is, changes that lead to promoting the competitiveness of the studied unit. The broken lines are undesirable changes which are not pursued through innovation. They lead to negative changes in the SWOT table and cause deterioration in competitiveness. The primary objective of SWOT in this dynamic pattern is to promote the competitiveness of the studied unit (e.g., company/industry) through innovation.

Creating innovation strategy and developing innovation activity, requires focusing on all areas of SWOT. The relation of these four 'boxes' to innovation is different in each case. The main goal is to develop the existing strengths and to create new ones in order to be more

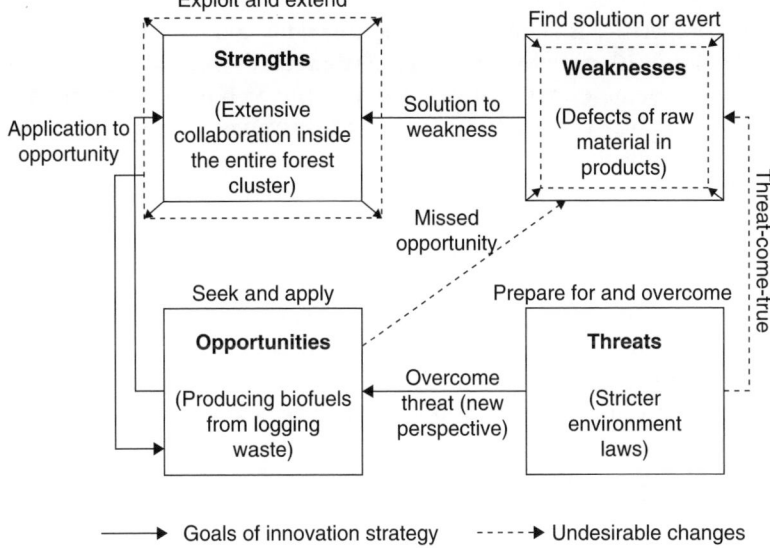

Figure 6.2 Applying SWOT in innovation strategy (case Finnish forest industry)

competitive. On the other side of the coin are the weaknesses to avoid and minimize their effects on competitiveness. When looking at the external environment, innovation is aimed at responding to the identified opportunities, shifting them to the 'strengths –box', as well as at defending one against the threats and possibly even shifting them to the 'opportunities box'. The balance between these boxes is case-dependent and has to be considered in relation to the factors identified in each box. Because some of the factors are likely to be more significant and likely to come to fruition than others, they require different focus and actions in the strategy process. Nevertheless, the target is to downsize the negative side of the SWOT (i.e., W and T) and to foster the positive side of it (i.e., S and O).

Dynamics of ... strengths

The internal strengths of a firm support innovation and sustain competitiveness, providing a competitive edge compared to rivalries. A strength can be a resource that the competitors do not have or it can be a way to produce a product with lower costs. It can be a renewal raw material or a production process with more advanced technology. Strength can also be a core competence of a firm. This is not the case in every 'S', because a strength can be understood as a single skill or technology, whereas a core competence is defined as a bundle of skills and technologies.[5] The aim of innovation strategy is to exploit strengths as efficiently as possible and to create new ones, that is, expanding the 'S –box'. As strengths are sustaining and even increasing the competitiveness, it is important to have strengths that have larger effects on the competitiveness. These are the factors that separate a firm from its competitors. Related to innovation, this means that a firm has to exploit these strengths in developing innovation and also develop innovations that nurture the strengths so that the competitive edge will remain. The worst-case scenario related to strength is a situation where a strength shifts to a weakness. This can be a result of either market-based or technology-based change and in some cases is impossible to prevent. This transition leads to decreasing competitiveness and needs actions that are defined below, where weaknesses of SWOT are dealt with.

In the case of Finnish forest industry, one of the strengths is extensive collaboration in the entire forest cluster. This means that firms in the forest industry have partners throughout the value chain closely located, having the same working culture and world cutting-edge technology. Forest industry companies benefit from extensive collaboration with other industries in Finland in developing innovation, especially process innovations, which in most of the cases have their origin in companies developing paper machines and chemicals needed in papermaking. New, high-tech process innovations advance the production technology of Finnish forest industry companies and thus give them competitive advantage compared to the foreign peers. The implication for innovation strategy is that it should exploit this strength in the best possible way in order to exploit the most recent R&D results throughout the value chain in developing new technology.

... Weaknesses

The internal weaknesses are the negative factors that reduce or destroy competitiveness. The aim of innovation is to find a solution to a weakness, or, if this is not possible or rational, to avert a weakness so that its effects to the competitiveness are reduced. It is not always wise to start developing a solution through an innovation to a weakness because of the risks and cost involved. Then the focus should be on averting the weakness, in other words trying to minimize its effects on competitiveness. This can happen, for example, by developing new products which do not suffer from the weakness as much as previous products. However, in many cases the pay-back of innovation will cover the resources put into the development of the technology solving the problem. In these cases it is reasonable to direct the innovation strategy to getting rid of the weakness. It should be noted that the benefits from an innovation as well as from a solved weakness can be hard to measure because they are often qualitative and not only quantitative, that is, not measurable only in money or in market share.

In the Finnish forest industry, one weakness is the defects of the raw material. Because wood is an organic substance, its dimensions change along with the air humidity. As compared to companies producing rival products from plastic, for example, wood processing companies cannot compete with the physical and chemical stability of the raw material. Some FFI companies have developed ways to minimize this problem by changing the wood reactivity to water. Innovations like antiseptic treatment and heat treatment have reduced the changes in wood dimensions, as well as the rotting of wood. These examples have enabled companies to avert the weakness by innovating new products which do not suffer as much from the weakness as did the previous products. A solution to this problem emerged in 2006, when UPM-Kymmene innovated a product called ProFi Deck – planks. It combines recycled wood and plastic to produce a material that does not react with water. With the radically new product the firm shifted the weakness to strength, increasing its competitiveness through innovation and creation of differentiated product. As this example shows, one innovation can decrease the number (and also effect) of weaknesses and increase the number (and also effect) of strengths.

... Opportunities

The opportunities box in SWOT is the one relating to most changes through innovation. Factors identified as opportunities are the positive external issues having potential to become strengths. The goal in innovation strategy is to seek these potentials and apply them when possible and beneficial. Opportunity can be open in the present environment or it can be a future change which is seen as likely to occur. The most beneficial timing in applying an opportunity is not always as soon as it is possible to do. As mentioned by Hamel and Prahalad (1994), the goal is not to be the first in absolute sense, for example, introducing a new product, but to be the first one with the product to finally unlock the emerging market. It can also be more profitable to seize on an opportunity after a while, with lower cost and risk, and higher return. For instance, it is not wise to seize a technical opportunity to develop a radically new product before the technology is advanced to a level where the final product meets the market needs and performance requirements and is easy enough to adopt and use.[6] Rather than being the first to grasp every opportunity, it is more important to get to the best 'O's. This might require medium-term strategy, for instance, in the form of a new product, in planning the pathway to taking full advantage of the most rewarding opportunity, for instance, an emerging new product segment.[7] Applying an opportunity through innovation shifts the certain opportunity to strength and thus increases competitiveness, but it can also create new opportunities in the environment. For instance, developing a new radical product innovation may well lead to a possibility of incremental innovation in the future, as this new product can be improved through customer feedback and new technical possibilities.[8] Foreseeing the impacts of taken opportunity, that is, future rewards and risks, is often hard and thus the long-term innovation strategy cannot be static but a dynamic one driven by the evolving external and internal situation.

Another change related to the 'O box' is an opportunity which is missed. This can happen when a new possibility is not identified or not applied by the firm, but is applied by its competitors, giving them competitive advantage. Thus, an opportunity becomes a weakness for a firm, because the technological development in the external environment advances but the firm lags behind in the technological trajectory, or because it decides not to follow a certain trajectory at all as a strategic choice.[9] This can have severe consequences, for example, in a situation where a competitor is the first to patent a crucial technology, possibly leading to a new product generation. In a worst-case scenario, the competitor might be able to put forward competence-destroying technological discontinuity that leads to decreasing competitiveness, or even to the exit of the firm that is not able to apply the same opportunity (see Tushman and Anderson, 1986).

A present opportunity in the Finnish forest industry is to produce new end products from the same raw material, for instance, biofuel from logging waste. One of the biggest firms, Stora Enso, announced in 2007 plans to apply this opportunity.[10] This will open up totally new markets for the company in the field of vehicle fuels, assuming that an economically sustainable biofuel product can be developed and commercialized, of course. As this application shifts the opportunity to strength, it also loops backwards, creating new opportunities in the form of developing new possible fuel mixtures and other chemicals from logging residue. Of course, applying an opportunity of this scale, or any other opportunity, is not risk-free and thus it should not be assumed that the right thing always is to grasp an opportunity when possible; the risks and rewards related to it should be considered before making the strategic decision. It should be mentioned that in the case of Finnish forest industry the position in the traditional product segments[11] is becoming less competitive and forcing enterprises to seek and apply ever more risky opportunities. This also shows that one single factor of SWOT should not lead to action in innovation strategy on its own, but the strategy should be build on a wider picture, taking into account all the factors (i.e., position of the firm and future changes). Firms need to follow competitors' reactions to emerging opportunities when making decisions, that is, stay up–to–date in the present external situation. Every opportunity is not a chance to take, but one to evaluate in the context of firm's strategic position and its future changes.

... Threats

Threats are the factors that can endanger competitiveness by becoming weaknesses. They might also have, at least in some cases, potential for opportunity. After identifying possible threats, firms need to prepare for them, and, if they become reality, to overcome them. Again, the risks, costs and benefits need to be studied case by case before taking action. If a threat, identified or unidentified, turns into reality it shifts to weakness, reducing the competitiveness. There are many reasons why this could happen. A firm might not have been able to identify the threat and the new situation of the business environment is a total surprise. Another explanation can be that the probability of a threat to become a reality is undervalued, or firm is, for some reason, unable to prepare for the threat at hand. If a threat is on that level that affects also the peers of the firm, it can also provide a positive possibility, or even a strength, when overcome. This is the case when a firm is able to treat it from a different point of view and turn it into an opportunity, while the rivals are unable to do the same.[12] For instance, the rapid development of China and the growing production capacity there creates threats for European companies in many industries. Some companies have been able to change the angle in this dilemma, and have used the benefits of growing markets and cheaper production costs of China by shifting some of their production capacity there.

For the companies in the Finnish forest industry, much debated, stricter environment laws are a threat. In 2008, this threat became a reality as a result of laws relating to emission trading which meant that every factory has a certain limit for its carbon dioxide emissions. If this limit is exceeded, the factory has to buy a licence for extra emissions or pay a fine for releasing too much carbon dioxide into the air. A factory or a company can also 'sell' its emissions capacity if it believes that it will not need as high a limit as it has, that it will run short of its yearly emission limit. In an energy-intensive sector like forest industry, this has meant a threat of losing competitiveness compared to sectors that can use less emissive energy sources or do not depend as much on energy. Some forest firms have been able to change their perspective on this threat and prepared for it. By developing new innovations, for example, emission filters, they have changed the threat into an opportunity and by applying these innovations, they have managed to change the opportunity, furthermore, to strength. Hence, innovating and preparing for and overcoming a threat has created a new strength, which the multinational companies can exploit even globally in several of their factories. There are also companies which have struggled with the threat of emission costs being realized and have been unable to answer to it. If these companies are not able to reduce their emissions under the limit which they will otherwise exceed, this threat will turn into a weakness. This will increase their costs of production compared to their competitors and put them into a weaker position to start with. As a conclusion, it can be mentioned that some threats can be turned into opportunities. In these cases the negative future prediction can turn out to be a positive one, forcing companies to innovate and stand out from their rivals.

Concluding discussion

A rapidly changing environment requires firms to put more focus on continuous change. As the need for innovation and renewal increases, so does the requisite of a dynamic innovation strategy to steer the innovation activity for long-term success. Formulating and reshaping a successful innovation strategy enables companies to be agile and competitive. To achieve this, firms need to be more proactive in strategy-making and to understand the interaction of innovation and the competitive situation of the firm. This is because innovation and technological development changes the internal characteristics of a company in relation to its peers, and/or the external environment it operates in. This strategic position in turn acts as an input for making innovation strategy, framing future innovations. By understanding the relationship of innovation, strategic position and innovation strategy, it will be possible to create and implement a successful innovation strategy proactively in response to the opportunities and threats of tomorrow in order to stay ahead of competitors.

This chapter has presented a pattern of the dynamics in strategic position related to innovation, applying SWOT, and how it can be used in planning and implementing innovation strategy. The points made in the chapter aim at helping firms to understand the nature of changes that technological development generates in the strategic position and support firms to implement the analysis to more continuous planning instead of a situational analysis. With this, the benefits of using SWOT or other strategy tools can be more than traditionally believed in companies.

The results show that, in creating and modifying the strategy, different aspects of SWOT and their importance need to be taken into account. However, different aspects of SWOT tend to have peculiar relationships with innovation. Building and enhancing strengths, as well as reducing the impact of weaknesses, can be seen as related to incremental innovation. The 'S's and 'W's are the factors that support firms in their strategies in the short term in order to be faster, better and more efficient than their competitors. Contrary to these, applying new emerging opportunities and overcoming threats is more related to radical innovation. They are the factors that open up the possibility to be different and stand out from the competitors more explicitly. The 'O's and 'T's give the possibility to be successful in the long run but require the company to be proactive as well as dynamic in developing and implementing its innovation strategy.

The study also raised the issue of managerial implications for firms in creating a strategy to become even more competitive. The biggest challenges for present-day companies in rapidly evolving, or even transforming, industries (like the Finnish forest industry, for instance) is to find a balance in their strategy. By this I mean that there is a need for both innovation strategy perspectives; incremental strategy and radical strategy. The first one is the 'easy' part of the process. It is based on the present-day situation and its obvious short-term trends, having targets nearby and easily reachable. It is about trial and error, testing and adjusting, monitoring and improving. There are no big risks and no big rewards. It will keep the firm in the markets, but it will not take it to new, emerging markets.

The other side of coin is much more demanding in terms of having the knowledge to create longer-term, radical innovation strategy and having the guts and risk-taking ability to implement it. The far future is much harder to foresee, while the strength-building to seize the most rewarding opportunity is no straightforward manoeuvre either. Yet, in order to be at the frontier when the new product segment is possible to develop and/or the mega-market opens, one needs to have a long-range plan identifying the possible big opportunity in time and the ways to grasp it. This will need freedom from prejudice to realize these big chances or even to create them.

The challenge is in creating a combination of incremental and radical innovation strategy, driven by the dynamic external environment and internal factors supporting and hindering innovation, in order to stay competitive and create even bigger dynamics by developing radical innovation. This, in turn, in terms of radical or transformative innovation, emerges as a new challenge in the form of a novel and fast-changing technological path on which to stay. Thus, a key to success is to comprehend the link between innovation and the firm's strategic position in formulating strategy.

Limitations

SWOT is not the only possible, or necessarily the best possible tool of strategic management. This chapter concentrates only on applying SWOT in innovation strategy and does not consider other ways to formulate and shape strategy. It should also be mentioned that the supposed changes in SWOT relate to innovation and technological change when the tool is used in innovation strategy. The dynamics of SWOT might be totally different when used in other kinds of strategy process, for example, market strategy.

The dynamics of innovation and strategic position is, at best, only a general picture of the changes taking place in it and takes no stand on the qualitative aspect of the identified factors. It displays the routes of factors, illustrated using the SWOT model, affected by innovation, but does not tell us which factors or characteristics of strategic position are the ones requiring primary focus in innovation strategy. Each SWOT is different, case by case, and thus it is a firm-specific (and industry-specific) question to look at the significance of each factor.

Appendix: SWOTs of the Finnish forest industry

SWOT in 1991 (Finnish Forest Industries Federation, 1990)

STRENGTHS	WEAKNESSES
High production capacities and advanced technology	Capital costs high because of recession
Operating efficiency through mergers and acquisitions	High levels of emissions and environmental wastes
Strong in main markets in Europe	High debt to turnover ratio
Self-sufficiency in raw material	
High profitability	

(Continued)

OPPORTUNITIES	THREATS
Integration across the value chain	Production capacity exceeding the global demand
Global operations	Increasing demand, and price level, of electricity
Growth through investments	Recession lowering profitability dramatically
Growing European markets	

SWOT in 2006 (Pesonen, 2006)

STRENGTHS	WEAKNESSES
Renewable raw material	High usage of energy, thus high costs from it
End products absorb carbon dioxide	High price level of domestic wood
Extensive collaboration in the entire forest cluster	High staff expenses
Economies of scale in case of multinationals	Geographical distance from main, and growing markets
High quality of products and production process	Physical and chemical defects of raw material
Finnish education system and capable employees	Commercializing innovations in mechanical wood industry
State-of-the-art technology	Custom tariffs for Russian wood
Sustainable forestry	Cheap dollar against euro
	Released carbon dioxide of decomposed/ burned end products

OPPORTUNITIES	THREATS
New knowledge from collaboration is co-funded	Increasing energy prices
Technology programs	Getting enough raw material to factories
New, radical products or even product segments	Changing customer needs
Faster and cheaper production process	Changing environmental laws and regulations
Improved innovation process	Environmental consciousness affecting the image

Notes

1. Mintzberg (1994) calls them internal and external audits, while Hamel and Prahalad (1994) speak of intellectual leadership in a broader sense, which includes most importantly foreseeing the future opportunities and evaluating the core competences of the firm. Porter (1980) included personal values in internal evaluation.
2. Meaning that the environment a) remains unchanged, b) is easily predictable or c) acquiesces to organization's strategy.
3. Adapted from Finnish Forest Industries Federation (1992) and Finnish Forest Industries Federation (2007).
4. By easy, I mean that SWOT is quite simple to understand and to put into use. By applicable, I refer to the high suitability of SWOT in different situations, i.e., it can be used in any kind of organization, industry, nation, etc., and with regards to different aspects or a part of strategy, e.g., marketing strategy.
5. See Hamel and Prahalad (1994), pp. 202. With their definition of core competence, it can be regarded in this case as a unique strength or strengths that is complex in nature, e.g., a capability of developing a modular process line having the possibility to produce new product innovations with few adjustments made. As a core competence is a bundle of skills and technologies, it can also be a bundle of strengths. Core competence is treated in somewhat the same way as strengths in strategy creation as they give a company the ability to exploit an opportunity and enhance competitiveness.
6. For instance, the first banking services through special software in mobile phones did not attract the markets, since they were more complicated to use than traditional internet banking services.
7. Hamel and Prahalad (1994) speak of strategic architecture as a broad opportunity approach plan which tackles the question of what must be done today, in terms of competence acquisition, to prepare to be the leader in an emerging opportunity arena. In other words, it is important to think how to use strengths in order to get to the most desirable opportunities and what strengths are needed in addition to the present ones. Teece *et al.* (1997) called this ability to achieve new forms of competitive advantage 'dynamic capability', which has to be built, not bought.
8. For example, the case of computer hard drives, which have improved continuously after the first breakthrough.
9. A good example of this is Nokia, with its decision to stay out of 'clamshell' phones in the 1990s. First mass-market phones using this flip phone technology and design by Nokia were commercialized in 2004, whereas Motorola released the first clamshell mobile phone, called StarTAC, in 1996.
10. In this consortia were VTT Technical Research Centre of Finland and Neste Oil, in addition to Stora Enso. They decided to build a small-scale factory for developing the technology further, towards a process of fuel production from wood-based biomass.
11. Pulp and paper, timber and plywood.
12. Effective managers should not be considered as the ones who avoid all the crises, but moreover those who exploit in opportunistic ways the crisis they know they can not avoid (Mintzberg, 1994)

References

Allio, R. J. (2006) 'Strategic Thinking: The Ten Big Ideas', *Strategy & Leadership*, Vol. 34, No. 4, pp. 4–13.

Ansoff, H. I. (1984) *Implanting Strategic Management*, Englewood Cliffs, NJ: Prentice Hall.

Chiesa, V. and Manzini, R. (1998) 'Towards a Framework for Dynamic Technology Strategy', *Technology Analysis & Strategic Management*, Vol. 10, No. 1. pp. 111–29.

Clarke, K. and Thomas, H. (1990) 'Technological Change and Strategy Formulation', in R. Loveridge and M. Pitt, *The Strategic Management of Technological Innovation*, West Sussex: John Wiley, pp. 273–91.

Cohen, W. M. and Levinthal, D. A. (1990) 'Absorptive Capacity: A New Perspective on Learning and Innovation', *Administrative Science Quarterly*, Vol. 35, No. 1, pp. 128–52.

Dosi, G. (1982) 'Technological Paradigms and Technological Trajectories: A Suggested Interpretation of the Determinants and Directions of Technical Change', *Research Policy*, Vol. 11, pp. 147–62.

Dosi, G. (1984) *Technical Change and Industrial Transformation: The Theory and an Application to the Semiconductor Industry*, London: Macmillan.

Finnish Forest Industries Federation (1990) *The Key to the Finnish Forest Industry*, Helsinki: FFIF.

Finnish Forest Industries Federation (1992) *The Key to the Finnish Forest Industry*, Helsinki: FFIF.

Finnish Forest Industries Federation (2007) *Basic Statistics of the Finnish Forest Industry* [web document] [updated 14 June 2007], available at:' http://www.forestindustries.fi/Infokortit/Statistic per cent20service/Documents/ FinnishForestIndustry2007_english.pdf

Fusfeld, A. R. (1978) 'How to Put Technology into Corporate Planning' (originally published in *Technology Review*, 1978), in R. A. Burgelman, M. A. Maidique and S. C. Wheelwright (1996), *Strategic Management of Technology and Innovation* (2nd edn), Homewood, IL: Irwin, pp. 60–4.

Hamel, G. and Prahalad, C. K. (1994) *Competing for the Future*, Boston: Harvard Business School Press.

Hill, T. and Westbrook, R. (1997) 'SWOT Analysis: It's Time for a Product Recall', *Long Range Planning*, Vol. 30, No. 1, pp. 46–52.

Itami, H. and Numagami, T. (1992) 'Dynamic Interaction between Strategy and Technology', *Strategic Management Journal*, Vol. 13, pp. 119–35.

Markides, C. C. (1999) 'A Dynamic View of Strategy', *Sloan Management Review*, Vol. 40, No. 3, pp. 55–63.

Mintzberg, H. (1990) 'The Design School: Reconsidering the Basic Premises of Strategic Management', *Strategic Management Journal*, Vol. 11, No. 3, pp. 171–95.

Mintzberg, H. (1994) *The Rise and Fall of Strategic Planning*, New Jersey: Prentice Hall.

Nelson, R. R. and Winter, S. G. (1982) *An Evolutionary Theory of Economic Change*, Cambridge, MA: Belknap Press of Harvard University Press.

OECD (2005) *Oslo Manual: Guidelines for Collecting and Interpreting Innovation Data* (3rd edn), Paris: OECD and Eurostat.

Pesonen, P. (2006) *Innovation Management and its Effect in Forest Industry* (in Finnish), Espoo: VTT Publications 622.

Porter, Michael E. (1980) *Competitive Strategy: Techniques for Analyzing Industries and Competitors*, New York: The Free Press.

Porter, M. E. (1991) 'Towards a Dynamic Theory of Strategy', *Strategic Management Journal*, Vol. 12, Special Issue: Fundamental Research Issues in Strategy and Economics, pp. 95–117.

Reinmoeller, P. (2002) 'Dynamic Contexts for Innovation Strategy: Utilizing Customer Knowledge', *Design Management Journal*, No. 2, pp. 37–50.

Stenfors, S. and Tanner, L. (2007) 'High-level Decision Support in Companies: Where is the Support for Creativity and Innovation?', in S. Stenfors, *Strategy Tools and Strategy Toys: Management Tools in Strategy Work*, Helsinki School of Economics, A-297, Helsinki: HSE Print.

Teece, D. J., Pisano, G. and Shuen, A. (1997) 'Dynamic Capabilities and Strategic Management', *Strategic Management Journal*, Vol. 18, No. 7, pp. 509–33.

Tidd, J., Bessant, J. and Pavitt, K. (2001) *Managing Innovation: Integrating Technological, Market and Organizational Change* (2nd edn), Chichester: John Wiley.

Tushman, M. L. and Anderson, P. (1986) 'Technological Discontinuities and Organizational Environments', *Administrative Science Quarterly*, Vol. 31, No. 3, pp. 439–65.

7
Spatial Changes in Innovation Processes Over Time in Finland

Jani Saarinen and Juha Oksanen

Introduction

In recent years, there has been a growing interest in innovation, entrepreneurship and technological change, and their impact on regional and national economic development and welfare. It is generally accepted that innovation is a major, if not the most important, source of productivity growth and that R&D is also very important in this respect. In the new economic geography, the spatial aspect of industrial locations and innovative behaviour has been taken into agenda. A lot of research has been devoted to study agglomeration, urbanization and localization benefits of the countries and regions (Fujita *et al.*,1999; Ottaviano and Puga, 1998; Krugman, 1991a; Marshall, 1920). In another research tradition, the concept accessibility has been used in order to analyze the evolution of the regions (Hirschman, 1958; Myrdal, 1957). In some studies, such abstract concepts as 'production milieu' (Davelaar, 1991) and 'innovative milieu' (Camagni, 1991) have been used to highlight the importance of agglomeration and local networks. However, when it comes to knowing in precise detail the interconnections between geography, innovation and evolution over time, we get less clear answers.

In this chapter the geographical evolution of Finnish innovations and innovative firms is examined. Our goals are to provide new information on the geographical distribution of innovations and innovative firms over time and on changes in innovation processes and characteristics of innovations during the period 1945–98. Considering the first objective, we analyze the distribution of innovations (divided into four areas of industry) by applying the central–periphery approach (Hoover, 1948;[1] Krugman, 1991b) to study the changes in innovative activity over time.

Here, we follow closely the conceptual setting, which has been developed by Davelaar (1991).

The point of departure in order to analyze changes in innovation processes and characteristics of innovations, and their reflections on economic development in general, is to start from the micro level, from individual innovations and the firms responsible for developing and commercializing them. With help of the existing databases on Finnish innovations commercialized during the period 1945–98, we are able to use qualitative data on innovations, cut down to the various categories related to innovations and their development processes. However, the biggest advantage of our innovation data is the possibility to study long-term changes in the relationships between various characteristics of innovations and the spatial changes in innovation processes in Finland.

The structure of this chapter is as follows. This introductory section is followed by the theoretical second section, where the most relevant theories in innovation literature are discussed and a theoretical framework is developed. The main focus is directed at theories, in which the spatial aspect of innovation activity and the evolution over time are taken into account. The third section presents the data used in this study. In the fourth section, main findings about the changes in innovation processes and characteristics of innovations in time-space context are presented. The concluding fifth section provides key insights from the study, as well as a recapitulation of the main findings.

Empirical and theoretical background

Agglomeration (static state) and characteristics of innovations

Over the years, various factors have affected to the establishment of new firms. Usually, new innovative firms have been established in locations, in which knowledge base and critical factors (see Arrow, 1962) for the survival of the firm activities have existed. Before the development of the public research infrastructure, closeness to a large existing company from the same field of industry was seen as important. Also, the availability of raw materials was one of the factors affecting decision-making of company managers. However, since the public research infrastructure has started systematically to build up, the newest technical and scientific information has become easily available also for the new firms across the country.

In a regional context, R&D conducted in firms is not the only way of enhancing innovativeness. High innovativeness also requires a suitable environment, infrastructure and cooperation within clusters of firms

(Stern *et al.*, 2000; Porter and Stern, 1999). In this context, the presence of other sectors that support the innovativeness of a particular sector is important (Porter, 1998).

Urbanization, agglomeration, localization and other benefits accruing from external economies form one of the main channels that transform regional balance within nations (Fujita *et al.*, 1999; Ottaviano and Puga, 1998; Krugman, 1991a; Marshall, 1920). The term 'agglomeration benefits' can be seen to comprise both urbanization and localization benefits. Urbanization benefits accrue from the presence of several actors and sectors in the same geographical area. Localization benefits refer to the utility of firms owing to the presence of other firms in the same industrial sector.

There is a long tradition of seeing the accessibility of regions as significant for economic development (Hirschman, 1958; Myrdal, 1957). Regions close to markets are better off than those located further away from centres. Accessibility in terms of high-quality connections (infrastructure) to centres alleviates the disadvantage of a peripheral location. Accessibility depends on the location of geographical areas with respect to markets and the state of the infrastructure. In other words, accessibility is a factor related to agglomeration, since large agglomerations tend to have high accessibility due to the size of their own markets (Huovari *et al.*, 2001).

Human capital is regarded as a crucial factor for economic growth in a modern knowledge-based society. In particular, human capital is at the heart of innovative behaviour, which is the source of technological progress. Groundbreaking innovations, in turn, usually take place at a higher intensity in large agglomerations than at the periphery (Kangasharju and Nijkaamp, 2001; Freeman, 1990). Finally, agglomerations tend to have high accessibility due to the size of their own markets and high-quality connections to other agglomerations.

Firms' innovative efforts do not proceed in isolation, but are supported by external sources of knowledge (Kline and Rosenberg, 1986; Dosi, 1988). Firms which are located close to these sources will enjoy relative advantages over more distant firms and consequently tend to have a higher innovative performance (Beaudry and Breschi, 2000). Significant sources of external knowledge are local universities and public research centres. By operating close to these sources of knowledge, inventors and firms in a specific industry have a greater likelihood of sharing the latest knowledge.

In Scandinavian countries, the geographical perspectives of innovation activities have recently become popular. In a paper by Jonsson *et al.*

(2000), the Swedish medicine-technology sector was studied. They found that the innovative activity was highly concentrated, as some 80 per cent of the creation of new products and processes originated from the five metropolitan and urban areas (Stockholm, Göteborg, Malmö, Uppsala and Halmstad). The manufacturing sector instead was found to be less concentrated than other industries within the medicine-technology sector. Similar central–periphery patterns are found for the manufacturing industry in Norway. For example, Wiig and Isaksen (1998) find a clear central–periphery pattern when measuring different Norwegian regions' share of firms with innovation costs and share of firms producing new or significantly altered products. Peripheral regions had a substantially lower share of both. Moreover, Asheim and Isaksen (1996) show that the costs associated with innovation of firms in central areas are mainly made up by (or are related to) R&D, while the same costs of firms in peripheral areas are, on the other hand, mostly constituted by trial production and production start-ups. This suggests that firms in central areas are more concerned with radical innovations, while firms in peripheral regions are skewed towards incremental innovations and tend to 'import and alter innovations from outside' (Asheim and Isaksen, 1996, p. 23).

The complexity issue in innovation literature has a relative short history (see, e.g., Kline, 1990; Miller *et al.*, 1995). On the one hand, it is believed that complexity is an important characteristic of innovations that should be captured in successful innovation studies. This is due to the fact that complex products and systems play a vital part in the modern economy. On the other hand, it has been hypothesized that the complexity of an innovation is correlated with the innovation process – especially with the competence base of the innovation.

Until now, there do not exist any studies in which complexity of innovations is analyzed in the geographical context. By this we mean that, to our knowledge, the relationship between the geographical location of the innovative firm and the level of complexity of the innovations has not been studied. In general, the term *complex* is used to reflect the number of customized components, the breadth of knowledge and skills required and the degree of new knowledge involved in production, as well as other critical product dimensions (Hobday *et al.*, 2000; Wang and von Tunzelmann, 2000). In addition, the complexity issue has been related to the increasing systemic nature of innovations. This means that innovations nowadays consist of a large number of different parts or technologies, which are successfully tied together. From discussions presented above, it follows that in order to develop complex innovations, firms must increasingly rely on external knowledge bases, and develop close collaboration linkages with knowledge providers such as research centres and universities. As the last-mentioned institutions are located in urban agglomerations and large cities, we assume that innovations originating from central areas are more complex in nature compared with innovations commercialized by firms located outside the centres.

Space-time innovation patterns (let's roll the system . . .)

In the previous section, the objective was to highlight some of the main characteristics, which connect together agglomeration benefits and innovative firms and innovations. The focus was on the static state, in which all activities take place without any movement in time. In this section, the aim is to roll the system, both in terms of space and time. This will result a framework for the rest of the chapter, where changes in innovation processes and characteristics of innovations are analyzed in a space-time context. We begin with some classical studies, moving towards the most recent research conducted in this area.

In his book, *The Location of Economic Activity*, Hoover (1948) presents the concept 'technical maturing of industries'. In Hoover's setting, when an industry is young and its problems unfamiliar, it is likely that they are located in central areas. As time goes by and the specific industry matures, decentralization of activities takes place.[2]

It ought to be recognized that innovation is a spatial-dynamic process. Initial locations of new technological systems are often somewhat arbitrary (Krugman, 1991a). According to Markusen (1987), the initial stage of a technological trajectory is concentrated in a very few locations. Although not all new production sectors settle at the place of major inventions, in principle firms tend to agglomerate near innovating firms, mainly because of the need for skilled labour and information. Davelaar (1991) argues that during the incubation phase, when major product innovations are made, the swarming process of (new Schumpeterian) firms is concentrated in central (usually urban) areas, because early innovations are more dependent on the urban 'milieu' than subsequent innovations.

Possibilities to significantly improve existing products decrease during the competition phase: product innovations become more marginal and process innovations take over. These later innovations do not demand so much from the urban production milieu as innovations in earlier stages of a technological trajectory, because further improvements are concentrated on existing products (which have been proven to be the most successful) and also on production processes of these goods. Therefore, the presence of basic knowledge-producing institutions is not of crucial importance any more during this phase. Furthermore, in the meantime, there has been ample opportunity for non-central areas to develop adequate innovation infrastructure to meet new demands (Kangasharju and Nijkaamp, 2001).

During the stagnation phase of innovative behaviour, peripheral (often rural) areas may even be in a favourable position. The relatively low number of innovations and the standardization of the products reduce the importance of local factors even further, while price competition as the source of competitiveness may then favour peripheral areas where factor prices may be expected to be lower than in central areas.[3]

Due to the general trend in industrial development, it is not difficult to realize how products and systems have become more technology intensive. This has led in its turn to an increase in complexity of new products. While technology intensiveness as well as the level of complexity has increased, the money to be spent on these products has exploded. In modern economies, complex technologies have become a cornerstone of the country's whole economy. It has been stated that, in 1995, complex technologies comprised over 80 per cent of the 30 most valuable world exports goods (Hobday *et al.*, 2000).

In the 1950s, a common opinion regarding the development times of new products was that they would experience an increase in the future, due, among other things, to an increase in complexity. However, some decades later, particularly in the 1990s, the importance of being a fast innovator emerged as an important factor for corporate strategic considerations (e.g., Rothwell, 1994). More emphasis has been put on the development work of new products in order to shorten the development times. It has been noticed that a company that develops high-quality products rapidly has several options that it may pursue. First, it may start a new product development project at the same time as its competitors, but introduce the product to the market much sooner. Alternatively, it may delay the beginning of a new development project in order to acquire better information about market developments, introducing its product at the same time as its competitors but bringing to market a product much better suited to the needs of its customers. Finally, it may use its resources to develop additional focused products that more closely meet the demands of specific customer niches and segments. These options can give companies an advantage in the markets, which are characterized by intensifying competition and the rapid rate of technological change (Wheelwright and Clark, 1992).

Hypothesis

In the previous sections, the main theoretical findings related to the relationship between characteristics of innovations and economic space are presented. On the basis of this discussion, we formulate a hypothesis to be tested with the help of Finnish innovation data. We have divided the hypothesis into two different groups. The first group (a) contains the hypothesis, which focuses more on the static state of the development while the second group (b) deals more with changes in innovation processes and characteristics of innovations in a space-time context. In other words, this means that main findings are summarized in this section. Later on, in the fourth section, the main findings presented here are analyzed with help of Finnish innovation data.

H1-a (static): During the early phase of the industry, most of the innovations originate from central areas (Davelaar, 1991).

H1-b (rolling): An industry produces more innovations in earlier phases of its life cycle (Kangasharju and Nijkamp, 2001).

H2-a (static): Radical innovations take place in centres (Asheim and Isaksen, 1996).

H2-b (rolling): During the incubation phase, innovations are more radical in their nature (Davelaar, 1991; Markusen, 1987).

H3-a (static): Young innovative firms are located in the centres.

H3-b (rolling): In the early life cycle of industry, small firms have a relative advantage (e.g., Schumpeter, 1965).

H4-a (static): Firms located in centres are developing more complex innovations.

H4-b (rolling): Complexity of innovations increases over time (e.g., Hobday *et al.*, 2000).

H5-a (static): Development times of innovations are shorter in centres (Lehner and Maier, 2001), thus giving a faster innovation rate.

H5-b (rolling): Development times of innovations are decreasing over time (e.g., Wheelwright and Clark, 1992).

The data

The Finnish innovation database

The data we use in this study originates from the Finnish innovation data for the period 1945–98. There are two different datasets used, SFINNO innovation data covering the period 1985–98 and historical innovation data H-inno for the years 1945–84. The innovation data is based on the so-called literature-based innovation output (LBIO) method (Palmberg *et al.*, 1999; Pentikäinen *et al.*, 2002). An advantage of this method is that it can trace the exact location where the innovation was developed. According to previous results of this type of studies, a remarkable regional concentration of new product announcements has been discovered when analyzing LBIO data (Feldman, 1994; Brouwer *et al.*, 1999), while a similar concentration was not visible in R&D data (Kleinknecht and Poot, 1992).

The innovation data used covers some 3100 innovations commercialized in Finland by Finnish companies. An innovation is understood here as invention that has been commercialized on the market by a business firm or the equivalent. For each innovation, there is information on the commercializing firm. This information includes entry, exit, geographical location, turnover, number of employees, patents and industrial classification (SIC) according to the main industrial sector of the firm. An innovative firm has been defined as a firm, which has developed and commercialized a new product – an innovation.

The SFINNO data covers mainly the period of 1985–98. The collection process included two different phases. First, the collection of basic data on the innovations, which included both data on innovation (name in Finnish, a brief description, year of commercialization, TOL 95/NACE industrial field and technological class (ICP)) and on the commercializing firm (firm identification code, name and address, TOL 95/NACE industrial field, number of employees, turnover and patents). Simultaneously with the reviewing and collecting process, the basic data on innovations and firms were included in the database. The second step was to send a questionnaire, in order to gather additional data on individual innovations that only the innovators can provide. The questionnaire included questions which made it possible to sample between different types of innovations and firms, industrial fields and technological fields. Furthermore, questions on the degree of novelty and commercial significance of innovations, origin and diffusion of the innovation, the time dimension of innovation processes, the role of public promotion and R&D collaboration were included. Usually, the innovators were asked to evaluate the importance of some particular factor which contributed to the development process of innovation. The scale ranged from not important, through minor importance, important, to great importance (0–3). This information has turned out to be significant in in-depth studies (Palmberg *et al.*, 1999).

H-inno database contains 1600 Finnish innovations from the period 1945–98 (Saarinen, 2005). In order to get the historical data comparable with the already existing data, the same variables, with some small modifications, were collected and included in the database. Hence, the main difference between the H-inno and the SFINNO datasets is the collecting process itself. In SFINNO, the so-called holistic approach was used, which means that different methods were used for identifying innovations (literature reviews and experts). Considering the H-inno data, only the literature based method was applied (technical literature and company histories).

In order to get some indications about the degree of novelty, the classification provided by OECD (1997) has been applied in Finnish innovation data. This means that the novelty of innovations has been analyzed from the perspective of the firm. Thus, an innovation is considered novel if it is new to the firm, and therefore has required some reconfiguration or accumulation of the knowledge base of the firm. To get some knowledge about the novelty of innovation in a geographical respect (or from the market perspective), a simple classification 'new to the Finnish markets' versus 'new to the world markets' has been included. This type of classification has also been implemented from the OECD's definition, in which the distinction was made between firm-only innovations and worldwide innovations. It is also compatible with evolutionary theories, which stress the complex set of interactions between innovation and dynamic competencies of the firm (Henderson and Clark, 1990; Teece, 1988; Teece *et al.*, 1994). In this particular chapter, in order to be radical, innovation has to fulfil two requirements: first, it has to be totally new to the commercializing firm and, second, it has to be new to the world markets.

Considering the complexity of innovations, we divided them into four classes of complexity. Our taxonomy originates from Kleinknecht *et al.* (1993), as they divided innovations into three classes by complexity. High complexity referred to a system comprising a large number of functional parts and coming from several different disciplines. Innovation of medium complexity in the Kleinknecht study was a unit of more than functional parts and required the integration of a few different disciplines. Innovation designated as being of low complexity required knowledge from one discipline and consisted of only one part. The underlying assumption was that the simpler the structure of an innovation, the less development is required. In our taxonomy, we have divided Kleinknecht's medium complexity innovations into two groups, depending on the complexity of development. Our four classes of complexity are the following: 1) High complexity: innovation is a system consisting of several functional parts, development is based on several disciplines; 2) Medium artefactual complexity/high developmental complexity: innovation is a unit, development is based on knowledge bases from several disciplines; 3) Medium artefactual complexity/low developmental complexity: innovation is a unit, development is based on knowledge base from one discipline; and 4) Low complexity: innovation is a single coherent unit. In this particular chapter, we have combined classes 1 and 2 together, and defined a high complex innovation as an innovation which belongs either of these classes.

In order to take part in the ongoing discussion on changes in development times, data on the years of major phases in the innovation's development cycle were collected. In the SFINNO survey data, the respondents were asked to indicate the years of major phases in the innovation's development cycle, including the year of basic idea, first prototype, commercialization, break-even point and first exports. In cases where commercialization or exports had not yet occurred, the respondent was asked to indicate this. The year of the basic idea is considered to indicate the year when the first initiative for the development of the innovation was voiced. The year of commercialization marks the year when the innovation entered the market on a larger scale than that of a mere prototype. Considering H-inno data, all data related to important phases in the innovation's development cycle are collected from the literature. Unfortunately, only in some 20 per cent of the innovations are development times known. In this context, the development time of an innovation is defined as the time it takes from basic idea to commercialization.

Division of regions

In this study, we have divided Finland into three different classes, following the methodology used in, for example, a study by Kangasharju and Nijkamp (2001). The cities and municipalities have been subdivided into central, intermediate and peripheral[4] classes on the basis of the GDP of the area. We have not only focused on the size of the central city of the area; we have considered, in addition, the surrounding sub-region (NUTS 4) as one region, which benefits from the presence of one large city. This implies that we expect the spatial diffusion to emerge not only according to physical distance to central regions, but rather according to their ability and willingness to adopt innovations (approximated here by size of a city).

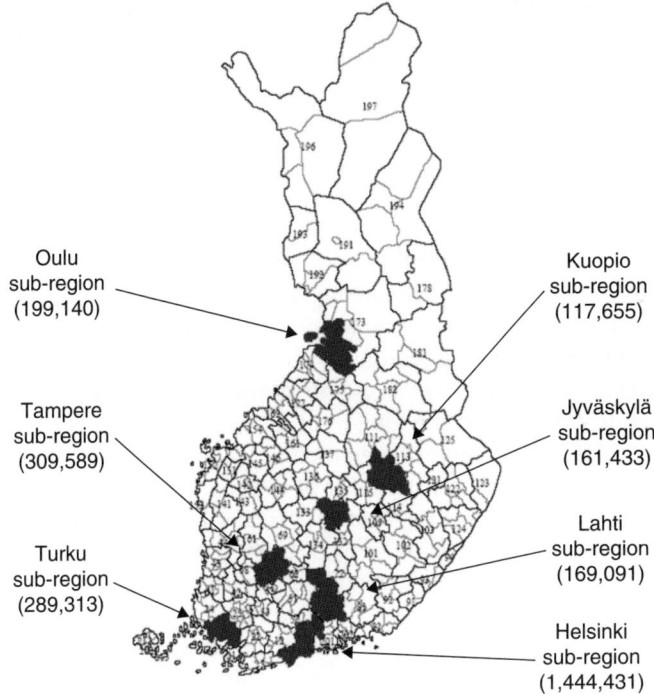

Oulu
sub-region
(199,140)

Kuopio
sub-region
(117,655)

Tampere
sub-region
(309,589)

Jyväskylä
sub-region
(161,433)

Turku
sub-region
(289,313)

Lahti
sub-region
(169,091)

Helsinki
sub-region
(1,444,431)

Figure 7.1 Map of Finland (number of inhabitants)

In this study, we have seven central areas and seven surrounding sub-regions. The central areas are Helsinki, Tampere, Turku, Oulu, Lahti, Jyväskylä and Kuopio. All of these cities are nowadays considered as 'growth centres', which means that they are growing more rapidly than the other areas in Finland, and they attract people to move in from the peripheral areas. The cities and villages belonging to our seven sub-regions are based on the latest division of geographical areas in Finland. This division was made in 2001, and it was taken into use immediately. The seven sub-regions we have included in this study are coloured in the map (Figure 7.1). A complete list of the cities belonging to the particular sub-regions can be found in Appendix 1. Areas which are not coloured in the map are considered as peripheries.

Altogether, if we calculate the number of inhabitants together, the central cities with surrounding sub-regions have 2.7 million people, which is more than half of the population of the whole of Finland. When we count the inhabitants living in our seven central cities together, the number is 1.3 million. The future expectations for our selected areas are that the population will continuously concentrate during the next decades. Worth noticing is that this growth is not caused by an increase in births. The largest part of this population comes from the uncoloured areas of the map, namely from peripheries.

Results

Overview of the Finnish innovation data

In order to get some idea about the coverage of the innovation data, some basic results are presented. Among all the variables collected, the year of commercialization is probably the best to start with, because this information is available for almost all of the innovations. It also gives some indications about the long-term development of the innovative pattern in Finland, in rather general terms. Figure 7.2 shows the number of innovations according to the year of commercialization.

The first observation from Figure 7.2 is that the general trend in the number of innovations is increasing over time. After some steady decades in the 1940s and 1950s, the number of innovations achieved an unstable and accelerating pattern towards the end of the studied period. During the period 1945–75, the number of innovations commercialized yearly varied, basically (with few exceptions), between 20 and 50 innovations per year. From 1975 until the early 1990s, between 50 and 90 innovations saw daylight yearly. Since the early 1990s, the number has increased rapidly, and far more than 100 innovations are introduced to the markets yearly by Finnish companies. Starting from 1996, the numbers decrease due to the lag in the rate at which innovations are reported

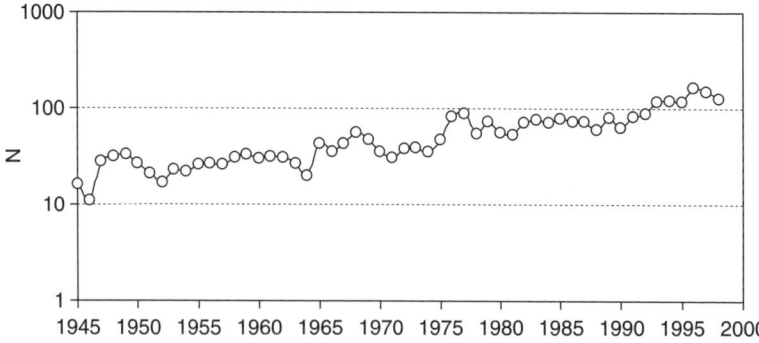

Figure 7.2 Number of innovations according to the year of commercialization

in the literature. For period 1945–84, 77 per cent of innovations were reported in the literature at the same year they were commercialized, 13 per cent were reported one to two years after the commercialization and 10 per cent after three years or more. For period 1985–98 the numbers are 43 per cent same year, 37 per cent one to two years after and 30 per cent three years or more after the commercialization.

The detected pattern in Figure 7.2 gives some clear indications about the increasing level of innovative activities of Finnish firms, but also about the fact that the method for identifying innovations is consistent from year to year. The trend also makes sense tentatively since it is consistent with other indicators, such as R&D expenditures of GDP and domestic patent applications.

The next step is to look at the geographical distribution of innovations. For that purpose, innovations have been divided into three different groups, according to the location of the commercializing firm.

The main message in Figure 7.3 is that the share of innovations originating from central areas, such as large cities, have decreased in 50 years from over 70 per cent to 30 per cent (see Appendix 4 for figures). Considering the industrial structure in Finland, as well as the relocation of heavy machinery-based industries to the less populated areas, the detected pattern is in line with expectations. However, if the geographical development should follow the theory of new and maturing industries (Hoover, 1948), the emergence of the electronics industry in the late 1970s, followed by the ICT boom, both known as really innovative industries, should be noticed from the figure. In order to get the

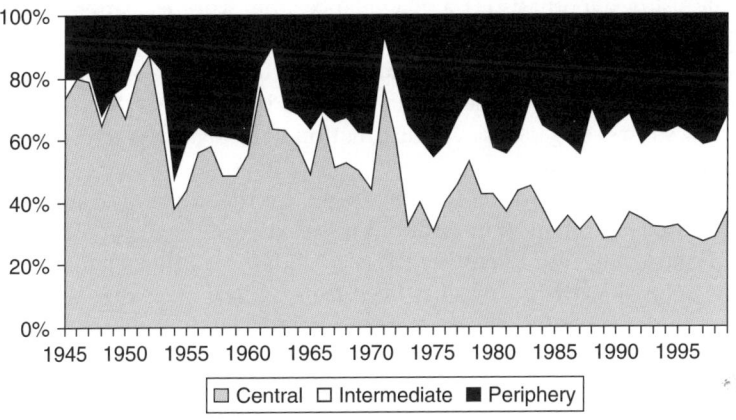

Figure 7.3 Geographical distribution of all innovations

pattern of new industries out of the figure, the development of these particular sectors has to be studied.

Considering the intermediate areas, their share of all innovations began to increase in the early 1970s; this trend continued over the subsequent decades. Here, one major explanatory factor has been the development of the cities Espoo and Vantaa, particularly Espoo's location close to Helsinki, as well as the moving of Helsinki Technical University from Helsinki to Espoo in the middle of the 20th century. Also the presence of Nokia's R&D departments (from the 1970s) has influenced to the pattern in Figure 7.3.

In peripheral areas, the long-term (ten-year) average has been in a slight increase since the beginning of the period. In the early 1950s, the share passed the 30 per cent level and, since then, it has varied between 30 and 40 per cent. Worth noticing is that, during the 1980s and 1990s, peripheral areas have been more innovative than central and intermediate areas, as measured by the share of commercialized innovations.

In order to get some indication about the distribution across different industrial sectors, we divide the whole industry into four different categories. The first category is called traditional industries, which includes branches like wood and paper products, metal, textiles, foodstuffs, vehicle and chemical industries (ISIC 11–28). Machinery industry consists of manufacturing of machines and machinery (ISIC 29). The electronics and electrical industries include manufacture of electrical and optical equipment (ISIC 30–3). In case of the software industry, telecommunication services (ISIC 64) and computer software and services (ISIC 72) form the base for this sector.

By using hypothesis H1-a as guidance, the expected pattern should be that during the early phase of the industry most of the innovations originate from central areas. We examine the four groups of industries, by taking into account the developments both in time and space. Table 7.1 below shows the geographical distribution of innovations across various industries over time.

The most interesting result from the table is that the central areas have decreased their share of innovations irrespective of the sector concerned. This means that our hypothesis H1-a is valid. In traditional and machinery industries, the change has been from central areas directly to the peripheries, whereas in electronics and electrical industries, the intermediate areas have advanced greatly. Considering the software industry, the central and intermediate areas are competing on the top position.

Although our first hypothesis turned out to be correct, the second hypothesis H1-b is a little bit trickier. According to H1-b, an industry

Table 7.1 Geographical distribution of innovations across various industries

	Traditional industries			Machinery industries			Electronics & electrical			ICT industries		
	Central (%)	Intermed. (%)	Periphery (%)	Central (%)	Intermed. (%)	Periphery (%)	Central (%)	Intermed. (%)	Periphery (%)	Central (%)	Intermed. (%)	Periphery (%)
1945–54	63	5	32	67	7	26	91	7	2			
1955–64	45	15	40	61	6	33	78	10	12			
1965–74	50	13	37	53	11	36	54	30	17	33	67	0
1975–84	40	20	40	39	17	44	50	30	20	41	35	24
1985–94	31	23	46	36	25	39	30	38	33	30	42	27
1995–	26	23	52	34	18	49	27	49	24	37	38	24

produces more innovations in earlier phases of its life cycle. As was seen already in Figure 7.2, the number of innovations in our database is increasing continuously. This is the case also in our four industrial sectors. The graphical illustration of the development of innovations divided into four industries can be found in Appendix 2a. Appendix 2b shows the share of innovations originating from central, intermediate and peripheral areas. A general observation from the figures presented in Appendix 2b is that in traditional, machinery and electrical and electronics industries the number of innovations originating from the central areas has more or less been at a constant level (with some minor variations) over the studied period. In traditional and machinery industries the biggest increase in number of innovations has been experienced by peripheral areas, whereas in the electrical and electronics industry, intermediate areas have advanced mostly. In the software sector, the main drivers behind the increase in innovations have been central and intermediate areas. Altogether, it can be concluded on the basis of the figures in Appendix 2a that all industrial sectors are continuously producing more innovations, irrespective of the evolution phase of the industry.

Characteristics of innovations: static state

In this section, we systematically go through the hypothesis related to the static state of innovations and innovation processes. Due to the limitations and incompleteness of the data, the results presented here are based on different numbers of innovations. For instance, considering Tables 7.3 and 7.6, the innovation data from period 1985–98 is based on the survey results, not on the whole stock. In addition, the low number in Table 7.6 relates to the collection process of historical innovation data, in which the development times for all innovations were not mentioned in the scanned magazines. Despite these difficulties with the data, the results presented here might give some new insights into the geographical characteristics of innovations and changes in innovation processes. We begin with radical innovations in the static state, and construct the following table.

Table 7.2 shows the distribution of radical innovations across geographical location. Only radical innovations are included in the table. For those readers who are interested to see the size of the share of innovations in various industrial sectors classified as radical, we refer to Appendix 5. In this section, our aim is to study how well the hypothesis H2-a fits with the Finnish innovation data. According to the hypothesis, radical innovations should take place in centres. If we look only at the total average numbers, the hypothesis seems to be valid, as almost

Table 7.2 Distribution of radical innovations according to geographical location (N = 2317)

	Central (%)	Intermediate (%)	Periphery (%)
Traditional industry	33.0	25.6	41.5
Machinery industry	41.8	17.6	40.6
Electrical & electronics	49.0	29.0	21.9
Software industry	43.1	33.8	23.1
Total average	*44.9*	*19.5*	*34.3*

Table 7.3 Distribution of young firms (0–9 years) according to their geographical location (N = 3107)

	Central (%)	Intermediate (%)	Periphery (%)
Traditional industry	32.8	19.1	48.1
Machinery industry	39.0	19.8	41.2
Electronics & electrical	41.1	31.0	27.9
Software industry	42.0	29.6	28.4
Total average	*37.4*	*23.2*	*39.4*

45 per cent of all radical innovations are commercialized by firms located in central areas. Even at the sectoral level, the hypothesis works well, traditional industry being the only exception.

In Table 7.3 the distribution of young firms across the geographical areas is presented. Hypothesis H3-a was that young innovative firms are located in centres. If we look at the average numbers for whole industry, the central areas are not achieving the highest shares. In fact, the largest share of young innovative firms is located in peripheries. This is quite an interesting result in the sense that it is against theories on agglomeration benefits, in which the local factors should encourage the innovativeness of the region and the easy entry of young innovative firms. In addition, the low share of young firms located at the intermediate areas runs in the face of this. In rapidly growing cities like Espoo and Vantaa (both located around the Helsinki), as well as growing municipalities in to the vicinity of Turku, Tampere and Oulu, the continuous construction of science parks and business incubators should have some impact on Table 7.3. In relatively new industries such as electrical and electronics, as well as in the software industry, the share of young innovative firms is higher compared to traditional and machinery industries. However, they are surprisingly close to the shares of peripheries.

Table 7.4   Distribution of high-complexity innovations according to geographical location (N = 3068)

	Central (%)	Intermediate (%)	Periphery (%)
Traditional industry	43.5	26.9	29.0
Machinery industry	48.9	14.5	36.0
Electrical & electronics	48.0	20.0	18.7
Software industry	37.1	35.3	27.6
Others	16.7	50.0	16.7
Total average	*44.8*	*19.4*	*34.2*

When continuing at the industrial sector level, there are some sector-specific patterns to be noticed. In sectors where the evolution has jumped directly from incubation to stagnation phase, most of the young innovative firms are located in peripheral areas. However, in the machinery industry, the difference between central and periphery is not very big. In this particular sector, 39 per cent of young firms are located in central areas, compared to 41 per cent in peripheries. In traditional industries, the distinction between central and peripheral areas is more notable. These results might indicate that as industry reaches the phase of stagnation, products become more specialized niche types of goods with high additional value, which have only a limited group of customers. The production facilities are relatively small, and the amount of produced goods is modest.

In electronics and electrical, as well as in the software industry, the pattern is slightly different. The largest share of innovations originates from the young firms located in central areas. In addition, the intermediate areas are also ahead of the peripheries. These results have some similarities with the patterns detected in Table 7.1. Industrial sectors in which innovations originate from central and intermediate areas seem to behave differently as compared with sectors in which central and peripheral areas are the main sources of innovations. In order to illustrate how complexity of innovations varies between different geographical areas, Table 7.4 is presented.

In hypothesis H4-a, which was developed in order to tackle the issue of complexity, we assumed that firms located in centres are developing more complex innovations compared with firms located in intermediates and peripheries. A look at Table 7.4 supports our hypothesis. In total, 45 per cent of high-complexity innovations are commercialized by firms located in central areas. In traditional, machinery and electrical and electronics industries the difference between central and other areas is quite

Table 7.5 Average development times of innovations in various industrial sectors across geographical location (N = 913)

	Central	Intermediate	Periphery
Traditional industry	3.6	4.9	3.3
Machinery industry	3.1	4.1	3.2
Electrical & electronic	4.0	4.1	2.9
Software industry	2.7	3.8	2.9
Others	6.0	6.6	4.5
Total average	*3.5*	*4.4*	*3.2*

significant. However, in the software sector, the intermediate areas are following closely behind. This pattern in the software industry might be explained by the nature of industry itself. The duties can be performed wherever computers are available. Due to the rapid growth of mobile solutions, geography is loosing its importance in this particular industry. In addition, emergence of new sector has provided an opportunity to broaden the industrial base in the region.[5]

As can be seen in Table 7.5, there are some great variations in the development times of innovations across geographical location. Overall, the longest development times are experienced in intermediate areas, where the average of all innovations is 4.4 years. This is almost one year more than in central areas and over a year longer than in peripheries. Irrespective of the industrial sector, the longest development times take place in intermediate areas. One explanation for this type of pattern might be that firms located in intermediate areas do not have such a close collaboration with research centres and universities (see Davelaar, 1991) as companies located in central areas. This means that, in order to be in the leading edge in the competition, innovative firms have to devote more time and resources to their in-house R&D activities. This is a time-consuming operation with an uncertain outcome.

Next, we turn our focus to the short development times. In terms of total average, peripheries have the shortest development times. At sectoral level, development times are shortest in traditional and electrical and electronics industries. In addition, the development times are equal[6] across different sectors. As our hypothesis H5-a was that development times of innovations are shorter in centres where the renewal rate is fastest, the results we get from the Finnish innovation data do not completely support this hypothesis. Only in machinery and software industries are the central development times slightly under those in the periphery. To get some reasonable answers for these results, we take the

Table 7.6   Phases of evolution

	Incubation phase	Competition phase	Stagnation phase
Traditional industry	1945–72		1973–98
Machinery industry	1945–72		1973–98
Electrical & electronics	1945–85	1986–98	
Software industry	1970–86	1987–98	

time aspect into account, and try to identify some space-time patterns of innovative activity of Finnish firms.

Characteristics of innovations: space-time patterns

In order to study space-time changes in innovation processes and characteristics of innovations, we apply Davelaar's (1991) model of industrial life cycle. In his model, Davelaar makes a distinction between three phases of technological change in spatial analysis, namely the incubation phase, the competition phase and the stagnation phase. During the incubation phase, central areas perform an above-average product and process innovativeness. Over time, the central areas become less and less innovative relative to the others, because intermediate and peripheral areas take over. The process innovativeness in central areas is lower than the product innovativeness, particularly after the incubation phase, because process innovations are realized more frequently during later phases and are therefore more concentrated on non-central areas.

On the basis of annual data of innovations and their geographical distribution, we are able to identify following phases.

Table 7.6 mainly strengthens the observation made already from Table 7.1, namely the different evolution paths of the industries. In traditional and machinery industry, there does not exist a period of time during which the largest share of innovations originates from intermediate areas. This means that these industries have transformed directly from the incubation stage to the stagnation phase. In addition, this change has taken place simultaneously. In electrical and electronics, as well as in the software industry, the evolution has proceeded according to the Davelaar's model, from incubation phase to competition phase.

What we do next is apply the division presented in Table 7.6 to combine space and time. In other words, we take the division into phases as granted, and combine different phases of evolution with each other. For instance, in the case of traditional industry, we compare the incubation and stagnation phases. In the software industry, the comparison is made between incubation and competition phases. In Table 7.7, the results

Table 7.7   Space–time patterns of innovations and innovative firms

	Share of Radical Innovations (%)	Share of young firms 0–9 years (%)	Share of high comp. innovations (%)	Average D-times of innovations
Traditional – Incubation	17.6	20.5	16.8	3.7
Traditional – Stagnation	36.7	32.2	20.9	3.8
Machinery – Incubation	12.8	20.7	21.6	3.5
Machinery – Stagnation	32.9	28.7	29.9	3.3
Electronics – Incubation	37.1	23.2	49.1	3.2
Electronics – Competition	60.0	60.9	73.9	4.1
Software – Incubation	48.9	28.9	73.3	2.8
Software – Competition	41.0	64.8	79.0	3.2
Others	62.1	13.8	41.4	4.2
Average Total	*31.8*	*29.3*	*33.1*	*3.6*

of our exercise are presented. With help of the table, we are able to get some answers to our remaining hypotheses, namely H2-b, H3-b, H4-b and H5-b.

Let's begin with hypothesis H2-b, which states that during the incubation phase innovations are more radical in their nature. In light of our results, this statement is valid only for the software industry. In this sector, 49 per cent of innovations are both totally new to the commercializing firm and are new to the world markets. This share is much higher as compared with the total average of 32 per cent. The highly radical nature of software innovations can partly be explained by the youth of the industry. With this we mean that, in Finland, the software industry emerged in the early 1970s. The period before that was mostly characterized by the low technological level of Finnish firms and innovations, particularly during the first decades after World War II (Dahmén, 1963; Saarinen, 2005). As the technological catch-up proceeded, particularly during the second half of the 20th century, Finnish products became more compatible in world markets. In the early 1970s, the level of engineering knowledge in Finland was already at an internationally compatible level. As a new industry saw daylight, Finnish industry was able, for the first time, to be among the innovators of the new software industry. In case of the other sectors, share of innovations originating from the incubation phase is lower compared with the forthcoming phases. For these industries, hypothesis H2-b does not work.[7]

Considering the next column, share of young firms, hypothesis H3-b emphasizes better opportunities for (small and) young firms during the early life cycle of industry. As can be noticed from Table 7.7, the results indicate something totally different. Interestingly, young innovative firms are not taking part of the development of an industry from the very beginning. According to our data, young firms enter markets after the incubation phase, particularly during the competition phase, as can be seen in the electronics and software industries. In these industries, more than 60 per cent of innovations are commercialized by young firms less than ten years old. This finding strikes against Schumpeter's early theory of economic development, in which young and small firms play a crucial role in the process of industrial renewal.

A general belief is that innovations become more complex as time goes by. This was tested with Finnish innovation data and, again, the hypothesis related to complexity H4-b is valid for all studied industries. If we compare the industries with each other, we see that over 70 per cent of all innovations from the software sector belong to the class of high complexity. On the other hand, only some 20 per cent of innovations originating from the traditional sector are characterized as high complex. Even this finding is in line with our common belief.

Finally, we focus on the development times of innovations. Hypothesis H5-b was that development times of innovations are continuously decreasing. In this respect, our findings behave rather differently. Solely in the machinery sector, the development times are decreasing over time. All the other sectors have witnessed an increase in development times. These findings get support from Saarinen's (2005) study, in which development times of innovations follow a cyclical pattern, not a decreasing one. The longest development times can be found in the electronics sector during the competition phase, whereas the shortest times belong to the incubation phase of the software industry.

Conclusions

As the results indicate, our studied period has witnessed some major changes. Considering the first goal of the study, the Finnish innovation data gives strong support to the central–periphery model. Innovations which are based on new and emerging technologies are commercialized by firms locating at the core of the province. As time goes by, new innovative companies are established in the intermediate and peripheral parts of the sub-region. As a result, the number of commercialized innovations becomes more evenly distributed between the geographical areas. Later on, peripheries take over the development and production of matured technologies, and a wave of innovations based on new emerging technologies are been commercialized by firms located at the core of the sub-region.

Table 7.8 Concluding table

	Traditional industry	Machinery industry	Electrical & electronics	Software industry
H1-a (static): During the early phase of the industry, most of the innovations originate from central areas	+	+	+	(−)
H1-b (rolling): An industry produces more innovations in earlier phases of its life cycle	−	−	−	−
H2-a (static): Radical innovations take place in centres.	−	+	+	+
H2-b (rolling): During the incubation phase, innovations are more radical in their nature	−	−	−	+
H3-a (static): Young innovative firms are located in the centres	−	−	+	+
H3-b (rolling): In the early life cycle of industry, small firms have a relative advantage	−	−	−	−
H4-a (static): Firms located in centres are developing more complex innovations	+	+	+	+
H4-b (rolling): Complexity of innovations increases over time	+	+	+	+
H5-a (static): Development times of innovations are shorter in centres	−	+	−	+
H5-b (rolling): Development times of innovations are decreasing over time	−	+	−	−

In order to provide an overview about the changes in innovation processes and characteristics of innovations in space-time context, we put our hypothesis together and present the following Table 7.8. As Table 7.8 shows, our findings with the Finnish innovation data do not completely support our hypothesis or the current findings in the literature. However, as the literature of innovation processes in space-time context seems to be virtually non-existent at the moment, there is no basis to draw too far reaching conclusions.

Appendix 1: List of central cities and their sub-regions in 2005

	No. of inhabitants
Helsinki	559,177
Tampere	201,980
Turku	175,059
Oulu	126,862
Lahti	98,349
Jyväskylä	82,409
Kuopio	90,000

Helsinki sub-region:
Espoo, Helsinki, Hyvinkää, Järvenpää, Kauniainen, Kerava, Kirkkonummi, Mäntsälä, Nurmijärvi, Pornainen, Siuntio, Tuusula, Vantaa

Tampere sub-region:
Kangasala, Lempäälä, Nokia, Pirkkala, Tampere, Vesilahti, Ylöjärvi

Turku sub-region:
Askainen, Kaarina, Lemu, Lieto, Masku, Merimasku, Naantali, Nousiainen, Paimio, Piikkiö, Raisio, Rusko, Rymättylä, Sauvo, Turku, Vahto, Velkua

Oulu sub-region:
Hailuoto, Haukipudas, Kempele, Kiiminki, Liminka, Lumijoki, Muhos, Oulu, Oulunsalo, Tyrnävä

Jyväskylä sub-region:
Jyväskylä, Jyväskylä mlk, Hankasalmi, Korpilahti, Laukaa, Muurame, Petäjävesi, Toivakka, Uurainen

Lahti sub-region:
Artjärvi, Asikkala, Hollola, Hämeenkoski, Kärkölä, Lahti, Nastola, Orimattila, Padasjoki

Kuopio sub-region:
Karttula, Kuopio, Maaninka, Siilinjärvi

Appendix 2a: Number of innovations in various industrial sectors

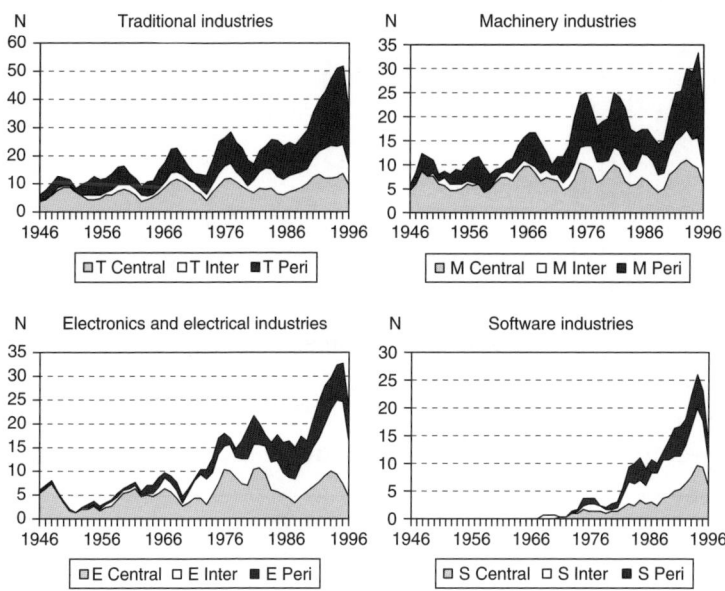

Appendix 2b: Share of innovations in various industrial sectors

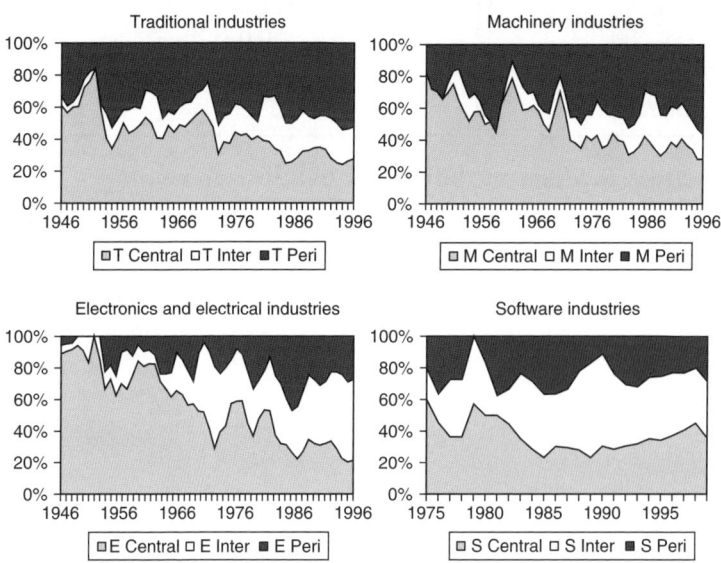

Appendix 3: Development times of innovations

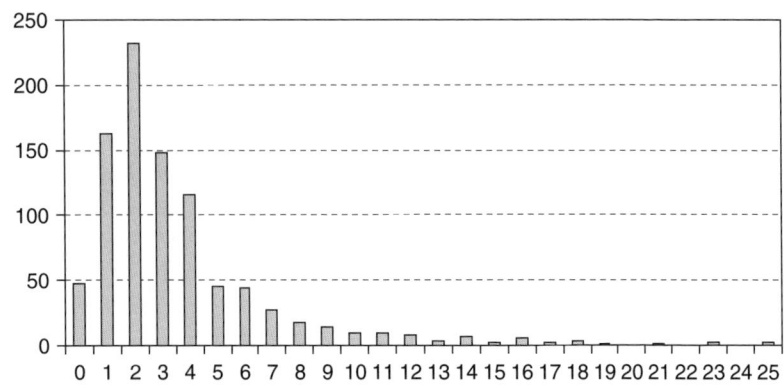

Note: Y-axis: Number of innovations
X-axis: Development time of innovation in years

Appendix 4: Geographical distribution of all innovations

	Central (%)	Intermediate (%)	Periphery (%)
1945–54	71.0	6.1	22.9
1955–64	57.1	10.8	32.2
1965–74	52.3	16.6	31.2
1975–84	41.8	21.8	36.3
1985–94	32.4	30.0	37.6
1995–	30.9	31.5	37.6
Total average	*41.6*	*23.2*	*35.2*

Appendix 5: Share of radical innovations from all innovations

	Central (%)	Intermediate (%)	Periphery (%)
Traditional industry	23.5	43.1	29.1
Machinery industry	21.5	30.9	26.0
Electrical & electronics	38.6	46.6	47.9
Software industry	50.0	39.3	39.5
Total average	*28.4*	*41.4*	*31.1*

Notes

1. Hoover used the concept 'decentralization with maturity' (see Hoover, 1948, pp. 174–6).
2. According to Keir's book on *Manufacturing* (1928), the locational histories of individual industries have very often involved an early stage of increasing concentration followed by a later stage of redispersion.
3. Compare with the ongoing globalization phenomenon.
4. Periphery classification includes also mid-size cities and towns (in Finnish terms cities with 30,000–80,000 inhabitants).
5. Municipalities also intentionally attract new sector firms through establishing technology parks etc.
6. If we do not take 'Others' into account.
7. One explanation could be dispersion of generic technologies into traditional industries. This has opened up new innovation possibilities.

References

Arrow, K. J. (1962) 'Economic Welfare and the Allocation of Resources to Invention', in R. Nelson (ed.), *The Rate and Direction of Inventive Activity: Economic and Social Factors*, Princeton, NJ: Princeton University Press.

Asheim, B. and Isaksen, A. (1996) 'Location, Agglomeration and Innovation: Towards Regional Innovation Systems in Norway', STEP Report R-13, Oslo.

Beaudry, C. and Breschi, S. (2000) 'Does clustering really help firms' innovate activities?', CESPRI WP n. 111, July.

Breschi, S. and Lissoni, F. (2001) 'Localised Knowledge Spillovers vs Innovative Milieux: Knowledge "Tacitness" Reconsidered', *Papers in Regional Science*, 80, pp. 255–73.

Brouwer, E., Budil-Nadrornikova, H. and Kleinknecht, A. (1999) 'Are Urban Agglomerations a Better Breeding Place for Product Innovation? An Analysis of New Product Announcements', *Regional Studies*, Vol 33, pp. 541–9.

Camagni, R. (1991) Technological Change, Uncertainty and Innovation Networks: Towards a Dynamic Theory of Economic Space', in R. Camagni (ed.), *Innovation Networks: Spatial Perspectives*, London: Belhaven-Pinter.

Dahmén, E. (1963) *Suomen taloudellinen kehitys ja talouspolitiikka. Vuodet 1949 – 1962 sekä tulevaisuuden näkymiä* (Ekonomisk utveckling och ekonomisk politik i Finland. En undersökning av åren 1949–1962 samt några framtidsperspektiv) (Finland's Economic Development and Economic Policy. Review of Years 1949–1962 and some Future Outlooks), Suomen Pankin Taloustieteellisen Tutkimuslaitoksen julkaisuja, Sarja C 4, Helsinki.

Davelaar, E. (1991) *Regional Economic Analysis of Innovation and Incubation*, Aldershot: Avebury .

Dosi, G. (1988) 'The Nature of the Innovative Process', in G. Dosi, C. Freeman, R. Nelson, G. Silverberg and L. Soete (eds), *Technical Change and Economic Theory*, London: Pinter, pp. 221–38.

Feldman, M. P. (1994) *The Geography of Innovation*, Kluwer: Doordrecht.

Freeman, C. (1990) *Economics of Innovation*, International Library of Critical Writings in Economics, 2, Brookfield: Edward Elgar.

Fujita, M., Krugman, P. and Venables, A. (1999) *Spatial Economy*, Cambridge, MA: MIT Press.

Henderson, R. and Clark, K. (1990) 'Architectural Innovation: The Reconfiguration of Existing Product Technologies and the Failure of Established Firms', *Administrative Science Quarterly*, Vol. 35, pp. 9–30.

Hirschman (1958) *The Strategy for Economic Development*, New Haven, CT: Yale University Press.

Hobday, M., Rush, H. and Tidd, J. (2000) 'Innovation in Complex Products and System', *Research Policy*, Vol. 29, pp. 793–804.

Hoover, E. M. (1948) *The Location of Economic Activity*, New York, Toronto, London: McGraw-Hill.

Huovari, J., Kangasharju, A. and Alanen, A. (2001) 'Constructing an Index for Regional Competitiveness', Pellervo Economic Research Institute Working Papers, No. 44, June.

Jonsson, O., Persson, H. and Silbersky, U. (2000) *Innovativitet och Regionen – Företag, Processer och Politik* (Innovativeness and the Region: Firms, Processes and Politics), Rapport 121, Swedish Institute for Regional Research, Östersand.

Kangasharju, A. and Nijkamp, P. (2001) 'Innovation Dynamics in Space: Local Actors and Local Factors', *Socio-Economic Planning Sciences*, Vol. 35, No. 1, pp. 31–56.

Keir, M. (1928) *Manufacturing*, New York: Ronald Press Company.

Kleinknecht, A. and Poot, T. P. (1992) 'Do Regions Matter for R&D?', *Regional Studies*, Vol. 26, pp. 221–32.

Kleinknecht, A., Reijnen, J. O. N. and Smits, W. (1993) 'Collecting Literature-based Innovation Output Indicators: The Experiences in the Netherlands', in A. Kleinknecht and D. Bain (eds), *New Concepts in Innovation Output Measurement*, London: Macmillan.

Kline, S. J. (1990) 'A Numerical Measure for the Complexity of Systems: The Concept and Some Implications', Report INN-5, Thermosciences Division, Department of Mechanical Engineering, Stanford University, CA.

Kline, S. and Rosenberg, N. (1986) 'An Overview of Innovation', in R. Landau and N. Rosenberg (eds), *The Positive Sum Strategy*, New York: National Academy Press, pp. 275–305.

Krugman, P. (1991a) *Geography and Trade*, Cambridge, MA: MIT Press.

Krugman, P. (1991b) 'Increasing Returns and Economic Geography', *Journal of Political Economy*, Vol. 99, pp. 483–99.

Lehner, P. and Maier, G. (2001) 'Does Space Finally Matter? The Position of New Economic Geography in Economic Journals', SRE-Discussion 2001/01, Department of Urban and Regional Development, Vienna.

Markusen, A. (1987) *Profit Cycles, Oligopoly, and Regional Development*, Cambridge, MA: MIT Press.

Marshall, A. (1920) *Principles of Economics*, London: Macmillan.

Miller, R., Hobday, M., Leroux-Demers, T. and Olleros, X. (1995) 'Innovation in Complex Systems Industries: The Case of Flight Simulation', *Industrial and Corporate Change*, Vol. 4, No. 2, pp. 363–400.

Myrdal, G. (1957) *Economic Theory and Underdeveloped Regions*, London: University Paperbacks.

OECD (1997) *Proposed Guidelines for Collecting and Interpreting Technological Innovation Data: 'The Oslo Manual'*, Paris: OECD and Eurostat.

Oksanen, J. (2003) 'VTT:n alueellinen rooli ja vaikuttavuus [Regional Role and Impact of VTT]', Espoo 2003, *VTT Tiedotteita – Research Notes* 2205.

Ottaviano, G. and Puga, D. (1998) 'Agglomeration in the Global Economy: A Survey of the "New Economic Georgaphy"', *The World Economy*, Vol. 21, No. 6, pp. 707–31.

Palmberg, C., Leppälahti, A., Lemola, T. and Toivanen, H. (1999) 'Towards a Better Understanding of Innovation and Industrial Renewal in Finland: A New Perspective', VTT – Technical Research Centre of Finland, Working paper 41/99, Espoo: VTT.

Pentikäinen, T., Palmberg, C., Hyvönen, J. and Saarinen, J. (2002) *Capturing Innovation and Recent Technological Change in Finland through Micro Data – Elaborating on the Object Approach*, VTT, Espoo.

Porter, M. (1998) *Competitive Advantage of Nations*, London: Macmillan.

Porter, M. and Stern, S. (1999) *The New Challenge to America's Prosperity: Findings from the Innovation Index*, Washington, DC: Council on Competitiveness.

Rothwell, R. (1994) 'Industrial Innovations: Success, Strategy, Trends', in M. Dodgson and R. Rothwell (eds), *The Handbook of Industrial Innovation*, Chelenham, Brookfield: Edward Elgar.

Saarinen, J. (2005) 'Innovations and Industrial Performance in Finland 1945–98', *Lund Studies in Economic History*, Vol. 34, Stockholm: Almqvist & Wiksell International.

Schumpeter, J. (1965) 'Economic Theory and Entrepreneurial History', in H. Aitken (ed.), *Explorations in Enterprise*, Cambridge, MA: Harvard University Press.

Stern, S., Porter, M. and Furman, J. (2000) 'The Determinants of National Innovative Capacity', NBER Working Paper 7876.

Teece, D. (1988) 'Technological Change and the Nature of the Firm', in G. Dosi, C. Freeman, R. Nelson, G. Silverberg and L. Soete (eds), *Technical Change and Economic Theory*, London: Pinter.

Teece, D., Pisano, G. and Shuen, A. (1994) 'Dynamic Capabilities and Strategic Management', working paper, University of California, Berkeley.

Wang, Q. and von Tunzelmann, N. (2000) 'Complexity and the Functions of the Firm: Breadth and Depth', *Research Policy*, Vol. 29, pp. 805–18.

Wheelwright, S. and Clark, K. (1992) *Revolutionising Product Development: Quantum Leaps in Speed, Efficiency, and Quality*, New York: The Free Press.

Wiig, H. and Isaksen, A. (1998) 'Innovation in Ultra-Peripheral Regions: The Case of Finnmark and Rural Areas in Norway', STEP Report R-02, Oslo.

8
Convergence in Innovation

Jani Saarinen and Robert van der Have

Introduction

In the literature of economics and innovation, the concept of convergence has been approached from a plurality of perspectives. Most commonly, convergence has been understood in technological terms or in the context of regional economic development within or between nations. In the past, the innovation concept, however, has not been linked explicitly to convergence. During the last decades, by contrast, innovation has emerged in the discussions related to national and sectoral systems of innovations, as well as in the growth and development policy of EU. Here, innovation has become very prominent as a means of convergence. Despite of the emerging role of innovation in these discussions, innovations and innovativeness has always been analyzed with help of some proxies (like patents, R&D investments, etc.). Characteristics of innovations and innovation processes have not been taken more systematically into account when new policies are planned and executed.

In this chapter, we approach the topic of convergence in innovation from various points of view. We begin by introducing convergence in national innovation systems and policies, as well as technological convergence between the sectors and the nations. Finally, we take a slightly different view on convergence, in which we analyze convergence at the level of individual innovations and innovation processes. That is to say, we analyze the process of convergence of the characteristics of innovations from various points of view. These include innovation process-specific variables such as the convergence in public funding of innovations, as well as innovation-related variables such as convergence in development times of innovations over time. The results of this chapter indicate that there are different types of convergence related to innovation, ranging all the way from macro (convergence in policy) to micro (convergence in innovation processes) level.

Convergence in national innovation systems/policy

To start with, we take a closer look at the convergence in innovation policies and national innovation systems. According to the traditional and widely used OECD definition (OECD, 1963), science and technology policy means the collective measures taken by a government to encourage the development of scientific and technical research and, on the other hand, to exploit the results of this research for general political objectives. In Finland, the construction and development of the instruments of technology policy began in the mid-1960s, later than in the large and more developed OECD countries (Lönnqvist and Nykänen, 1999). However, the late start was counterbalanced by a resolute and rapid catch-up. The national features of Finland's technology policy are based on the absolute and relative scarcity of resources, the dearth of military and other so-called big science and, until the late 1980s, the paucity of research-intensive industrial branches and firms that utilize and generate high technologies (Lemola, 2002b).

Finland's development strategy in science and technology policy has aimed at catching up with a high international level, or more precisely, with models of leading-edge countries and international organizations which, from the Finnish perspective, have been considered legitimate and successful (Lemola, 2002b). In Finland, Swedish practices and policies have been particularly influential, and the OECD has been the main vehicle for the international diffusion of information and ideas. When analyzing the history of Finnish science and technology policy in more detail, it is possible to divide it into three main lines, which have been developing parallel to each other. First, there is scientific research and its development; second, the development of university education and its administration; and, third, the connection between these factors or elements and industry (Lönnqvist and Nykänen, 1999).

According to Lemola, Finland's technology policy from the late 1960s onwards can be divided into three phases if analyzed from the policy perspective (Lemola, 2002b). The first phase, which began in the mid-1960s, is called the period of research policy. The emphasis was put on the construction of the instruments of research policy and its quantitative development. In order to improve the international competitiveness of Finnish firms, the government began to support firms' research and product development directly by means of R&D loans and grants. A new fund under the authority of the Bank of Finland, Sitra, was established for this purpose in 1967. Since then, Sitra's activities have expanded from the original task of financing technical research and development to cover a range of research, educational and venture capital activities that benefit the economy and society at large.

Another major public financier during the first phase was the Ministry of Trade and Industry (MTI), responsible for technology policy and providing support for industrial research and development. In addition, the MTI began to support the research and product development of firms. A motive of these measures was concern about the lack of firms' own R&D. Statistical studies made in the early 1960s showed that, with only a few exceptions, the level of private R&D was modest by international standards (Lemola, 2002b). The auspices of MTI contained, and still contain, a number of organizations such as publicly supported research institutes, agencies and state-owned companies engaged in special financing, which are an important part of the national innovation environment.

The focus during the second phase, which began at the turn of the 1980s, was on the development of technologies and is referred to as the period of technology policy. It was during that period the National Technology Agency (Tekes) was established and various technology programmes launched. Since its foundation in 1983, Tekes has been the major agency for the distribution of selective R&D support in Finland. Between 1985 and 1999, Tekes' R&D funding increased from 49–410 million euros. Tekes prepares, funds and coordinates national technology programmes, and provides funds for applied technical research and risk-carrying R&D ventures in industry. It also contributes to the preparation of national technology policy. Its mission is to promote the competitiveness of Finnish industry through the use of technological means. Moreover, Tekes has a regionally comprehensive organization that acts in conjunction with the Employment and Economic Development Centres.

The third phase, called innovation policy, began in the early 1990s. Innovation policy focuses on innovation-promoting factors and emphasizes aspects such as technology transfer, diffusion and commercialization (Lemola, 2003). In the late 1990s, policy-makers took an initiative to develop new models and tools in order to sustain the level of competitiveness of the Finnish economy in the future. An example is the use of the loosely defined concept of national innovation systems (NIS) (OECD, 1997) as a management tool in order to improve the competitiveness of Finnish economy (Science and Technology Policy Council, 2003). The results of this exercise are not discernible yet, or they are difficult to disentangle. However, in the European community of innovation policy-makers, the Finnish strategy has received a lot of positive feedback. In addition, it has become an object for a large number of imitative activities in various countries.

Research and development (R&D)

The empirical part for the innovation policy chapter comes from R&D. Research and development (R&D) is perhaps one of the most classical issues in studies concerning technological change and innovation activity of companies. Since the early 1960s, there has been no doubt in the economic literature about the importance of R&D activities in relation to the economic growth (Acs and Audretsch, 1988, 1991; Freeman and Soete, 1997; Griliches, 1995). One widely used measure of national investment in change-generating activities is the share of GDP spent on R&D activities. R&D expenditures are often seen as an input indicator in relation to the innovative process, whereas patents – and, later on, innovations – are regarded as an output indicator (OECD, 1986; Kodama, 1986). R&D expenditure reflects formal expenditure on research and development as reported to the Central Bureau of Statistics. Therefore, the main advantage of this indicator is that data are available on firm, industry and national levels (OECD, 1993).

According to Freeman and Soete (1997), the distinctive feature in modern industrial R&D is its scale, its scientific content and the extent of its professional specialization. The professionalization in its turn is associated with three main changes: (1) the increasingly scientific character of technology; (2) the growing complexity of technology; and (3) the general trend towards division of labour, as noted already by Adam Smith. Usually, the developments in the USA after World War II have been mentioned as 'golden years' in terms of the emergence of company R&D, as particularly the role of small and new firms in the commercialization of new technologies – computers, semiconductors and biotechnology – increased (Bruland and Mowery, 2005).

The first official statistics concerning scientific-technical activities were published in 1930s in the Soviet Union. A couple of decades later, in the 1950s, the Organization for European Economic Cooperation (OEEC) made an attempt to construct systematic tools to measure research activities. Since 1963, the objective has been to use internationally standardized statistical tools in order to measure R&D expenditures. In Finland, a systematic collection of R&D statistics was started by Statistics Finland in 1969. However, even before that, some attempts were made to measure the research expenditures in Finland (Törnudd, 1958; Elfvengren, 1958).

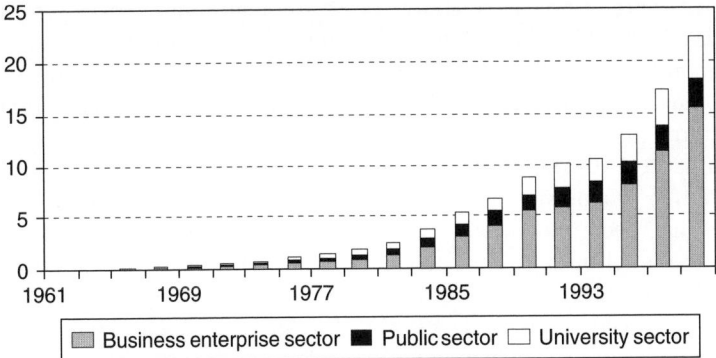

Figure 8.1 Long-term development of private and public R&D expenditures in Finland
Sources: State Council, 1974 (1962–72); Statistics Finland (1973–99)

Since the beginning of the period of research policy in the mid-1960s, R&D statistics have been one of the most observed indicators by Finnish policy-makers. New goals for the level of R&D expenditure have continuously been updated. As can be noticed from Figure 8.1, R&D expenditures have steadily increased. Not even the deep economic crisis in the early 1990s was able to hinder this increasing pattern. Since the early 1980s, industry has been the driving force behind the improvements of the level of R&D in Finland. The reason for this response was that Finnish products were not competitive enough in international markets. Simultaneously with the increase in private R&D expenditures, they have also become global. When exports began to advance, a new measure indicating the share of high-technology exports was taken into use. While the proportions between privately and publicly funded R&D were approximately 50:50 in the late 1970s, in 2000 the share was 70:30 to the advantage of private R&D. In the late 1990s, when the total R&D expenditures passed 3.0 per cent of GDP, the main emphasis was directed towards the increase of the level of public R&D spending. Today, the common goal for the EU countries is 3.0 per cent of GDP in 2010 ('The Lisbon Agreement'). This has resulted a massive process of convergence across the EU countries, as attempts to reach the 3.0 per cent level have been initiated.

Technological convergence (between countries)

After the innovation policy section, we continue our journey from macro level towards the micro level by discussing the technological convergence between the countries. Technological differences between countries have been a widely discussed issue during the last decades in economic history. There have been theories about the early and late adopters of technology (catch-up) (Denison, 1967; Abramovitz, 1986), national innovation systems (Freeman, 1987; Nelson, 1993), as well as competitive advantages of countries (Dosi *et al.*, 1990).

Perhaps the most common proxy for the technological level of countries has been the number of granted patents, both domestic and international ones. It has been stated that the US patent statistics can be considered the best source for assessing the patenting activity of any country outside the USA (Ray, 1988). More specialized research into these patents has usually helped us to understand both the level of technology in an international context of different countries and the countries' specialization in different technological fields. Despite some misleading qualities of patenting data, it is an important and informative variable, at least in long-term series studies. It has also been shown that small and medium-sized countries have a relatively high propensity to protect their innovations using patents taken out in foreign countries (Archibugi and Möller, 1993).

Patents

One of the most common ways to analyze the technological level between the countries is to use the patent data. In this section, we will introduce some general data about the US patenting activity of Sweden and Finland. When we have used the patent data, we have concentrated on the application year rather than the grant year. This has been done in order to avoid the time-lag which is some two years from date of application to the date of grant. In Figure 8.2, the development of US patenting of Sweden and Finland is introduced.

As Figure 8.2 illustrates, there exist quite big differences in US patenting between Sweden and Finland. After a rapid increase in the number of patents in the early part of the studied period, the level of Swedish patents has been stable, around 400 patents yearly. In the Finnish case, the number of patents started from a low level and has shown a clear increasing trend during most of the period. In per capita terms, Finland had almost reached the Swedish level by the late 1980s. During the last decade of the period, the development of the two countries' patenting has been rather similar. Both countries have experienced a rapid growth in US patenting, caused mainly by the electronic industries. Although these two countries are rather similar in their industrial structure, this

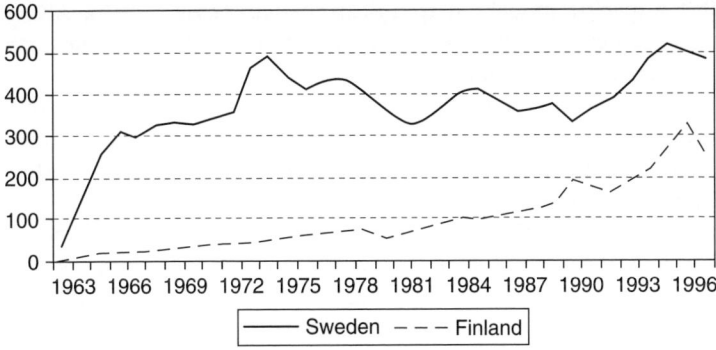

Figure 8.2 Total number of US patents

similarity at the aggregate level is not generally found at branch level (Lingärde and Saarinen, 2004).

Technological convergence (between sectors)

Differences across the industries and sectors have been a focus of large number of recent studies. The original idea of sectoral patterns of innovative activity was already developed by Schumpeter, as he emphasized the role of different sizes of firms and market structure as the underlying factors explaining sectoral differences (Schumpeter, 1911, 1942). Scherer, in studying inter-industrial differences in R&D investments and patenting, found that the most important determinants explaining sectoral differences were the scientific and technological opportunities, which vary largely among industries (Sherer, 1967). Nelson and Winter were interested in the environment in which the firm operates and the key characteristics of this environment. They defined a technological regime as 'a frontier of achievable capabilities, defined in the relevant economic dimensions, limited by physical, biological, and other constraints, given a broadly defined way of doing things' (Nelson and Winter, 1977, p. 57). Dosi, in turn, discussed the determinants and inter-sectoral patterns of investment in innovation and found factors like technological opportunities, appropriability conditions, knowledge bases and search procedures behind sectoral differences (Dosi, 1982, 1988). However, Dosi concluded that one model of technical change, suitable to describe the characteristics of all sectors, is simply not possible. Indeed, the peculiar characteristics of innovative processes historically observed in empirical studies brought Pavitt to the formulation of a taxonomy describing industry-specific models of technical change (Pavitt, 1984). Also the Finnish evidence displays clear differences between industrial sectors and, moreover, a pattern that has changed over time.

After World War II, the mechanical engineering and forest-based industries constituted the major pillars in the Finnish industrial life. Since the late 1960s, the Finnish science and technology policy has consistently been oriented towards upgrading the knowledge base of the country and putting a great emphasis on the significance of innovation and high technology (Lemola, 2002a). As a result, the electronics and electrical industries have entered the club of major sectors. During the 1990s, the goal was formulated to broaden the industrial base through the establishment of new firms, to encourage innovations in emerging fields, and to stimulate the diversification of older firms (Lemola, 2002a; Eela, 2001). As the industrial base has diversified during the last decades, it is of interest whether these changes have been reflected at the level of innovations. In Table 8.1, the distribution of innovations according to their product classes is presented. The numbers before the parentheses present the number of innovations in the particular product class, whereas the numbers inside the parentheses describe the share of innovations in the particular product class compared to the total number of innovations belonging to the same time period (as a percentage).

According to Table 8.1, a large majority of innovations belong to machinery and equipment and electro-technical products. It demonstrates the importance of the capital goods industry in general (i.e., machinery and capital equipment) as the main source of innovations for the entire production system. However, what has happened during the years is that machinery and equipment sector has lost rather rapidly its share of innovations in different cohorts (from 37.5 to 22 per cent), although it still was the largest group in 1985–98. Other sectors which have lost their relative share of all innovations are wood products, metal products and transport equipment. Although the transport industry flourished once in the 1950s and 1960s, it nowadays functions as a producer, not as a developer as it used to do. However, the transport sector has kept its share in output. Even electrical and electronics have slightly lost their share, which is a somewhat surprising result. When considering the success stories in terms of innovations, the biggest increase has been witnessed in the software sector. Software has already passed the electrical and electronics sector, being the second largest group of innovations. The other success stories have been foodstuffs, pulp and paper, chemicals, non-metallic mineral products and even textiles.

Table 8.1 The distribution of innovations according to product class

Product class	1945–66	1967–84	1985–98	Total N
Foodstuffs	4 (0.7%)	2 (0.2%)	81 (5.6%)	87
Textiles, clothing	4 (0.7%)	7 (0.7%)	17 (1.2%)	28
Wood products	47 (7.9%)	51 (5.1%)	34 (2.3%)	132
Pulp & paper products	10 (1.7%)	9 (0.9%)	53 (3.6%)	72
Oil & chemicals + pharmaceuticals	46 (7.8%)	81 (8.1%)	145 (10.0%)	272
Non-metallic mineral products	10 (1.7%)	23 (2.3%)	27 (1.8%)	60
Basic & fabricated metal products	56 (9.5%)	68 (6.8%)	109 (7.5%)	233
Machinery and equipment	222 (37.5%)	335 (33.5%)	321 (22.0%)	878
Electrical, electronics	72 (12.2%)	108 (10.8%)	169 (11.0%)	340
Instruments	41 (6.9%)	137 (13.7%)	160 (12.2%)	355
Transport equipment	68 (11.5%)	101 (10.1%)	44 (3.0%)	213
Electricity, gas & water supply	10 (1.7%)	39 (3.9%)	19 (1.3%)	68
Software	0	40 (4.0%)	221 (15.2%)	261
Others	2 (0.3%)	0	48 (3.3%)	46
Total	592	1001	1456	3049

Note: The numbers before the parentheses present 'number of innovations in the particular product class', whereas the numbers inside the parentheses describe the share of innovations in the particular product class compared to the total number of innovations belonging to the same time period

Source: SFINNO innovation data (see Appendix)

In a majority of LBIO–studies (note 1) the sectoral distribution of innovations is usually analyzed according to the industrial classification of the commercializing firm (i.e., the Dutch, Irish, US and Austrian studies). This kind of classification allows us to analyze the industrial structure of a country, in light of their innovating activity, more in detail. The negative side is that it gives little information about the distribution of innovations themselves. Not all innovations belong to the same class of industry as the firms responsible for their development. For instance, a firm located in the machinery sector might be responsible for development of innovations to a large number of different industrial sectors. Another negative side is that software innovations seem to be under-represented, in some studies even non-existent. The reason for this is that software development units have been located inside the manufacturing firms, as an integrated part of them. This was particularly the case still in the early 1990s (during the time when these above-mentioned studies were conducted), before the separate software sector emerged. In SFINNO data, innovations are classified both according to their product class and the industrial class of the firm. Fortunately, there do exist LBIO studies, in which the innovations are classified according to their product class. These studies include the Italian and the British study. In Table 8.2, the sectoral distribution of innovations from these three studies is reported.

As the correlation coefficients illustrate, the distribution of innovations according to their product classes is rather equal across the countries. In all three countries, a large share of innovations belong to

Table 8.2 International comparison according to innovation NACE

Product class	SFINNO (Basic) 1985–98	Ranking	The UK (II) study 1992	Ranking	The Italian study 1989	Ranking
Foodstuffs	81 (5.6%)	7	92 (9.8%)	4	38 (2.4%)	9
Textiles, clothing	17 (1.2%)	14	8 (0.9%)	13	49 (3.1%)	7
Wood products	34 (2.3%)	11	19 (2.0%)	9	17 (1.1%)	11
Pulp & paper products	53 (3.6%)	8	4 (0.4%)	14	43 (2.7%)	8
Oil & chemicals + pharmaceuticals	145 (10.0%)	5	88 (9.4%)	5	187 (11.7%)	4
Non-metallic mineral products	27 (1.9%)	12	14 (1.5%)	11	3 (0.2%)	13
Basic and fabricated metal products	109 (7.5%)	6	31 (3.3%)	8	38 (2.4%)	9
Machinery and equipment	321 (22.0%)	1	186 (19.8%)	2	439 (27.4%)	1
Electrical, electronics	169 (11.0%)	4	201 (21.4%)	1	201 (12.6%)	3
Instruments	160 (12.2%)	3	134 (14.3%)	3	155 (9.7%)	5
Transport equipment	44 (3.0%)	10	15 (1.6%)	10	67 (4.2%)	6
Electricity, gas & water supply	19 (1.3%)	13	14 (1.5%)	11	–	–
Software	221 (15.2%)	2	82 (8.7%)	6	350 (21.8%)	2
Others	48 (3.3%)	9	52 (5.5%)	7	15 (0.9%)	12
Total	**1456**		**940**		**1602**	

Correlations: SFINNO vs. UK = 0,830; UK vs. Italy = 0,756; Italy vs. SFINNO = 0,964 (All correlations are significant at the 1 percent level)

Note (1): The numbers before the parentheses present number of innovations in the particular product class, whereas the numbers inside the parentheses describe the share of innovations in the particular product class compared to the total number of innovations the machinery and equipment, and electrical and electronics sectors. The third class, which also takes a significant share of innovations, is the software sector. In Italy, more than 20 per cent of innovations are classified as software. Here, the UK is really lagging behind with its share less than 10 per cent. According to the results, the Finnish and Italian industrial structures, in terms of innovations, are quite similar (correlation 0.964), which is really a surprising result. By keeping in mind that the industrial structure in Italy has usually been associated as traditional, compared to the Finnish industry 'high-tech' label, the results might indicate two things. Either the structure on the Italian industry is not as traditional as it is thought to be, or the Finnish high-tech label is over-emphasized. Unfortunately, in lack of time-series data from other countries, we are not able to say whether there has been a process of convergence that has taken place or whether the industrial structure of these countries has been rather equal during the last decades. A wild guess would be that countries have come closer to each other as time has gone by.

In order to measure the sectoral distribution of innovation activities over time, the Herfindahl index is introduced. In the literature of economics, this index has been used to measure the concentration of different kinds of economic activities. This index allows us to answer the question whether innovation activity has been evenly distributed between the sectors or not. At each period in time this is given by

$$H = \sum p_i^2$$

where p_i is the sector's share of all innovations. If the index approaches zero, it can be stated that innovation activities are relatively evenly distributed across a large number of sectors. If the value of index approaches one, the innovation activity has been concentrated in a few sectors. In Figure 8.3, the Herfindahl index for the whole studied period is presented.

Figure 8.3 confirms the pattern shown in Table 8.1 that innovations are appearing more equally from different industrial sectors as time goes by. The observation goes also hand in hand with the patenting pattern of Finnish firms in the United States Patent and Trademark Office (USPTO). Since the beginning of the 1960s, the distribution of patents granted in the USA for Finnish applicants has become more even (Lingärde and Saarinen, 2004). On the basis of these results, innovations and patents, it can be stated that the Finnish industry nowadays is not dependent on only a few innovative sectors. Innovations are developed across the whole of industry very broadly speaking.

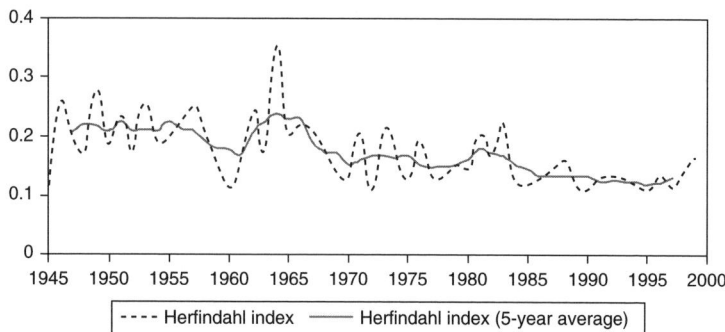

Figure 8.3   Herfindahl index according to the product classes of innovations

Convergence in innovation

In this chapter, the convergence in innovation processes and characteristics of innovation are analyzed with the help of two examples. Although a lack of comparable data from other countries makes the conclusions of this analysis less powerful, we still believe that it is a good starting point for a broader discussion. Particularly in the case of the role of public funding in innovation, the Finnish pattern differs largely from other countries' innovation policies. In the second example, in which we analyze the development time of innovations, the results might be more universal in nature.

Public support

In the economic literature, there exist a large number of studies in which the role of public intervention has been analyzed. In general these studies deal with the following hypotheses. First, that expenditures on R&D generate a positive effect on economic growth and social welfare (Mansfield, 1988; Griliches, 1984). A second hypothesis says that public intervention generates an additional growth of investments in R&D (Heijs, 2003). From this follows the third hypothesis that public support for R&D will have an additional positive effect on economic growth and social welfare. Of these three assumptions only the first one has achieved wide consent among the researchers in this field. Considering the last two assumptions, the discussion is lively.

In general, it is believed that the government has the responsibility to invest in the basic infrastructure of education and knowledge generation for science and technology, particularly in universities and research institutes. The educational system in its entirety is a public good that provides industry a skilled work force, citizens employment, parents child care and the public a cohesive society. There is a direct linkage between the amount of money given to universities and the number of graduates which will provide a skilled labour force for R&D.

In addition to investments in the education and training system, governments also provide financial support for the development of industrial processes and products. The Finnish government has used several modalities to allocate this money over a diverse set of science and technology actors, both in the public and the private sector, to generate technological knowledge. Previously, the money directed to that private sector took both the form of direct support to companies and of support for product development activities. The underlying idea was to promote the competitiveness of Finnish industry. Considering product development projects, it was assumed that by taking part of the R&D activities of firms, these would be more willing to develop new products for the market. If these products turn out to be successful, they will generate positive effects for the economy as a whole, for instance, in terms of employment. In case of direct subsidies for firms, the idea was to encourage firms with their main activities in strategically important technological areas. Such subsidies were forbidden when Finland joined the EU in 1995, due to the competition regulations. Since then, public funding takes place through support for R&D projects of firms.

A long-term objective for the Finnish technology policy has been to create a sound knowledge infrastructure and reduce the dependence on traditional sectors of economic activity, thereby transforming Finland to a knowledge-based economy. Since the 1990s, the fostering of new enterprises and innovations is the mainstay of Finnish technology policy (Science and Technology Policy Council of Finland, 1996, 2000). The tool which has been used to achieve this goal has been a selective R&D support, which is regarded as a means of promoting new business and enhancing the competitiveness of industry.

In Figure 8.4, we see the increasing share of innovations involving public funding with the 1980s as the most expansive period. Early attempts to increase the level of public funding had already taken place in the late 1960s. Despite the efforts made through the establishment of Sitra and the Foundation of Finnish Inventions, the share of publicly funded innovations largely did not increase during the 1970s.

During the period 1945–66, Finnish companies commercialized 592 innovations. Only five of them (0.8 per cent) involved some public funding. The funding organizations were Ministry of Trade and Industry

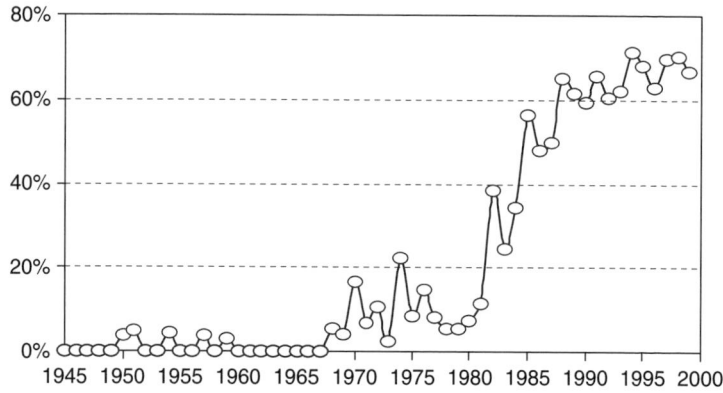

Figure 8.4   The share of innovations involving public funding, 1945–98

(1950), the Finnish Cultural Fund in two cases (1951 and 1959), Tekniikan Edistämissäätiö (note 2) (1954) and Bank of Finland (1957). After the 1950s, these organizations do not any more appear as financiers of innovations. An exception is Bank of Finland, which was the main financier of Sitra. However, Bank of Finland itself concentrated on other tasks, and did not directly fund R&D activities of single companies. Hence, it can be stated that before the mid-1960s the government support and promotion of companies was insignificant.

The 1970s witnessed some early attempts for a 'take-off' in terms of publicly funded R&D activities in private companies. Economic integration into Western Europe was set as the main goal of Finnish economic and social policy by the early 1960s. In the late 1960s, it was widely understood in Finland that in order to achieve economic and industrial growth, new innovations, patenting and R&D activities were needed. It was namely realized, in the light of an OECD report, that there was a lack of original, new and exportable products, for which foreign licences were not required. It was also assumed that there was a need to support industrial activity, particularly in sectors characterized as backward, but which had the potential for international competitiveness (Lönnqvist and Nykänen, 1999). In order to enhance economic growth, the only option for the Finnish industry was to adjust to international market integration.

The question now is: what is the optimum level of public intervention as it comes to development of innovations? In Finland, in some two-thirds of all industrial innovations public funding plays a significant role.

Certainly, public funding has helped many companies to achieve their dreams and develop new goods to the markets, but it might have some negative impacts as well. Until now, this policy has been successfully executed in Finland, but in the long run, do public authorities really have the best possible knowledge about the markets so that they are able to judge which technologies are worth developing and which are not? Should this type of selection belong to the companies, who operate frequently on tough international markets?

Development time of innovations

In the 1950s, a common opinion in the technical magazines considering the development time of new products was that they would experience an increase in the future. In the 1980s, it was recognized and widely accepted that there exist some large differences in development time of innovations in different industrial sectors. For sectors like software and traditional sectors, like foodstuffs and the forestry industry, the development times seem to be really short at the moment. In the case of software, the short development times probably reflect dynamic markets, in the sense that product life cycles are short, intellectual property rights are weak and firms need to innovate continuously (Toivanen, 2000). At present, there are sectors, like pharmaceuticals, which instead are characterized by extremely long – and still increasing – development times.

In order to analyze changes of the development time, data on the year of the basic idea and the commercialization were collected. The year of the basic idea is considered to indicate the year when the first initiative for the development of the innovation was voiced. The year of commercialization marks the year when the innovation entered the market on a larger scale than that of a mere prototype. In principle, the year of commercialization can be found for all included innovations from the articles in trade and technical journals, and the annual reports of the large firms. In cases where the written source material did not contain this information, the year of publication of the written source material was taken as a proxy for year of commercialization. In this context, the development time of an innovation has been defined as the time it takes from the basic idea to the commercialization. Figure 8.5 illustrates the average development time of innovations.

The most significant message in Figure 8.5 is that development time of innovations is not decreasing continuously over time. Instead, a cyclical pattern can be observed. As can be seen in the figure, the development time of innovations increases and reaches its maximum value during

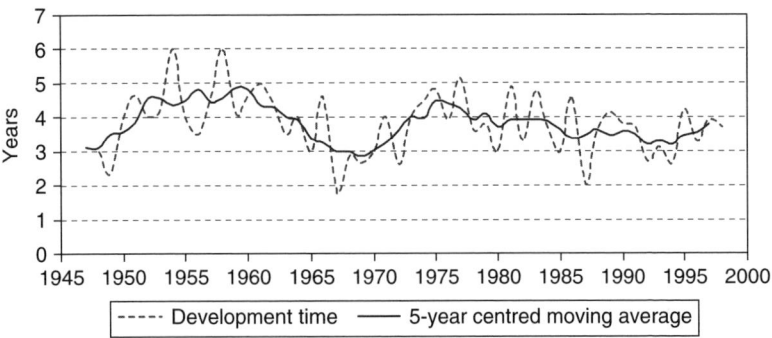

Figure 8.5   Development time of innovations

the fluid phases of the two technological eras. The first upswing, during the 1950s, was an era of mechanical engineering. A large number of innovations had their origins in machinery-based industries. In addition, these innovations were mechanical in their nature, consisting of pre-manufactured parts, which were joined together, resulting in new products – innovations – for new specific purposes. During the early years of an upswing, when the technologies are rather new, it takes more time to get the new products ready for the market. As the technology reaches its maturity, the number of radical innovations decreases and incremental innovations become more common. Simultaneously, there has taken place a learning process, resulting in a shortening of the development time.

While the first upswing in development time was dominated by mechanics, the second can be entitled an era of electronics, automation and instruments. This era had its fluid phase in the late 1960s, reaching its maturity phase in the early 1990s. A large number of innovations originating from this era are more or less mechanical in their nature, but they include electronic or automatic parts, which affect the way the product functions. Different types of knowledge are needed for the developing of new innovations during this era. One type of knowledge is needed in order to construct the mechanical frame for the product itself. Another type of knowledge is needed to join the electronics with mechanics, and to get it working. Before the technology has matured and its different elements become synchronized, the development time increases.

Can changes in development time of innovations be seen as an example of convergence in innovation, or would the right word be variance? There are a number of sectors, in which development times become shorter and shorter, approaching the development time of one week or even one day. On the other hand, sectors like pharmaceuticals experience longer development times than ever before. In view of a lack of comparable data from other countries, we will leave this question unanswered.

Conclusions

In this chapter, we have approached the issue of convergence in innovation from various points of view. We began by introducing convergence in national innovation systems and policies, technological convergence between the sectors and the nations, as well as the regional convergence. Finally, we took a slightly different view on convergence, in which we analyzed convergence at the level of individual innovations and innovation processes. The results of this chapter indicate that there are different types of convergence related to innovation, ranging all the way from macro (convergence in policy) to micro (convergence in innovation processes) level. As this chapter was the first of its kind, in which the issue of convergence is analyzed with the Finnish innovation data, it shoud serve as an excellent starting point for the future analysis of this interesting topic.

Appendix: SFINNO innovation data

The data we use in this section originates from the Finnish innovation data, SFINNO, for the period 1945–2005. The innovation data is based on the so-called literature-based innovation output method. The innovation data covers some 4500 innovations commercialized in Finland by Finnish companies. An innovation has been defined as an 'invention that has been commercialized on the market by a business firm or the equivalent'. For each innovation, there is also specific information on the commercializing firm (Saarinen, 2005).

SFINNO is a longitudinal database with its backbone consisting of some 4000 individual innovations, presently covering the years 1945–2005. The entire set entails basic data on these innovations and the firms commercializing them. This includes such variables as the product group of the innovation, the year of commercialization and the sector of the commercializing firm. In addition, survey data on the origins and diffusion of the innovation, aspects of R&D collaboration, public support and commercial significance has been acquired since the year 1985. The method of collecting the contemporary data (i.e., since 1985) is fairly 'holistic' in that it combines literature-based innovation counting, expert opinion, systematic reviewing of annual reports, as well as questionnaire instruments. The focus in the SFINNO database is on product innovations because study of in-house process innovations is not feasible with the object (i.e., innovation-centred) approach. SFINNO covers a wide variety of new products and sectors, varying from novel food items and methods for the preservation of wood, to the latest mobile technology, to name a few.

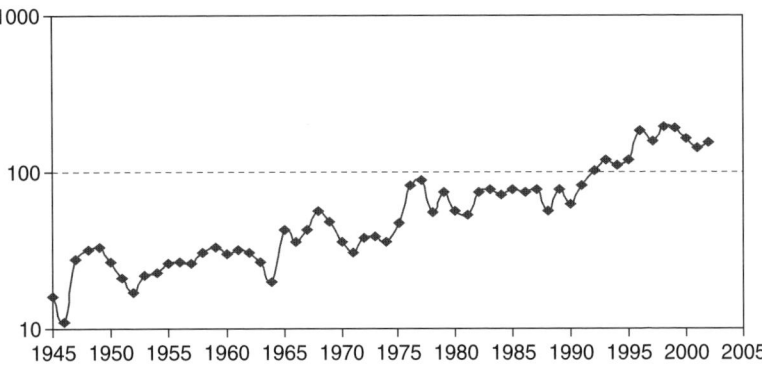

Number of innovations in the SFINNO data (log scale).

Notes

1. LBIO stands for literature-based innovation output study.
2. Tekniikan Edistämissäätiö means Technological Foundation.

References

Abramovitz, M. (1986) 'Catching Up, Forging Ahead and Falling Behind', *Journal of Economic History*, Vol. 46, No. 2.
Acs, Z. and Audretsch, D. (1988) 'Innovation in Large and Small Firms: An Empirical Analysis', *American Economic Review*, Vol. 78, No. 4, pp. 678–90.
Acs, Z. and Audretsch, D. (1991) 'R&D, Firm Size, and Innovative Activity', in Z. Acs and D. Audretsch (eds), *Innovation and Technological Change: An International Comparison*, Ann Arbor, MI: University of Michigan Press.
Archibugi, D. and Möller, K. (1993) 'Monitoring the Technological Performance of a Small Economy Using Patent Data: The Case of Denmark', *Technology Analysis & Strategic Management*, Vol. 5, No. 2.
Bruland, K. and Mowery, D. (2005) 'Innovation Through Time', in J. Fagerberg, D. Mowery and R. Nelson (eds), *The Oxford Handbook of Innovation*, Oxford: Oxford University Press.

Denison, E. D. (1967) *Why Growth Rates Differ*, Washington, DC: The Brookings Institution.

Dosi, G. (1982) 'Technological Paradigms and Technological Trajectories: A Suggested Interpretation of the Determinants and Directions of Technical Change', *Research Policy*, Vol. 11, pp. 147–62.

Dosi, G. (1988) 'Sources, Procedures and Microeconomic Effects of Innovation', *Journal of Economic Literature*, Vol. 26, pp. 1120–71.

Dosi, G., Pavitt, K. and Soete, L. (1990) *The Economics of Technical Change and International Trade*, Hemel Hempstead: Harvester Wheatsheaf.

Eela, R. (2001) 'Suomen teknologiapolitiikka valtion tiede- ja teknologianeuvoston katsausten valossa' (Finnish Technology Policy in the Light of the Reviews of the Science and Technology Policy Council), VTT, Group for Technology Studies, Working Papers 56, Espoo, 2001.

Elfvengren, E. (1958) *Kostnaderna för teknisk naturvetenskaplig och ekonomics forskning i Finland*, Taloudellinen Tutkimuskeskus, Helsinki: Sarja B, p. 10.

Freeman, C. (1987) *Technology Policy and Economic Performance: Lessons from Japan*, London: Pinter.

Freeman, C. and Soete, L. (1997) *The Economics of Industrial Innovation*, 3rd edn, London and Washington: Pinter.

Griliches, Z. (1984) 'Introduction', in Z. Griliches (ed.), *R&D, Patents and Productivity*, Chicago: Chicago University Press, pp. 1–19.

Griliches, Z. (1995) 'R&D and Productivity: Econometric Results and Measurement Issues', in P. Stoneman (ed.), *The Handbook of the Economics of Innovation and Technological Change*, Oxford: Blackwell Scientific.

Heijs, J. (2003) 'Freerider Behaviour and the Public Finance of R&D Activities in Enterprises: The Case of the Spanish Low Interest Credits for R&D', *Research Policy*, Vol. 32, No. 3, pp. 445–61.

Kodama, F. (1986) 'Technological Diversification of Japanese Industry', *Science*, Vol. 233, pp. 291–6.

Lemola, T. (2002a) 'Tiede- ja teknologiapolitiikan muotoutuminen', in Päiviö Tommila and Allan Tiitta (eds), *Suomen Tieteen Historia*, Helsinki: WSOY.

Lemola, T. (2002b) 'Convergence of National Science and Technology Policies: The Case of Finland', *Research Policy*, Vol. 31, pp. 1481–90.

Lemola, T. (2003) 'Innovation Policy in Finland', in P. Biegelbauer and S. Borrás (eds), *Innovation Policies in Europe and the US*, Ashgate: The New Agenda.

Lingärde, S. and Saarinen, J. (2004) 'Technological Specialisation in Sweden and Finland 1963–97: Contrasting Developments', in J. Ljungberg and J.-P. Smits (eds), *Technology and Human Capital in Historical Perspective*, London: Palgrave Macmillan, pp. 205–28.

Lönnqvist, K. and Nykänen, P. (1999) 'Teknologiapolitiikan alkuvaiheet Suomessa 1940–1970 luvuilla', VTT, Group for Technology Studies, Working Papers No. 40/99, VTT, Espoo.

Mansfield, E. (1988) 'The Speed and Cost of Industrial Innovation in Japan and United States: External vs Internal Technology', *Management Science*, Vol. 34, October, pp. 1157–68.

Nelson, R. and Winter, S. (1977) 'In Search of Useful Theory of Innovation', *Research Policy*, Vol. 6, No. 1, pp. 36–76.

Nelson, R. (ed.) (1993) *National Innovation Systems: A Comparative Analysis*, New York: Oxford University Press.

OECD (1963) *Science and the Policies of Governments: The Implications of Science and Technology for National and International Affairs*, OECD: Paris.

OECD (1986) *OECD Science and Technology Indicators*, OECD: Paris.

OECD (1993) *Frascati Manual*, Paris: OECD.

OECD (1997) *National Innovation Systems*, OECD: Paris.

Pavitt, K. (1984) 'Sectoral Patterns of Technical Change: Towards a Taxonomy and a Theory', *Research Policy*, Vol. 13, No. 6, pp. 343–73.

Ray, G. F. (1988) 'Finnish Patenting Activity', ETLA Discussion Papers No. 263, Helsinki.

Saarinen, J. (2005) 'Innovations and Industrial Performance in Finland 1945–98', *Lund Studies in Economic History*, Vol. 34, Stockholm: Almqvist & Wiksell International.

Scherer, F. M. (1967) 'Research and Development Resource Allocation Under Rivalry', *Quarterly Journal of Economics*, Vol. 81.

Schumpeter, J. (1911) *Theorie der wirtschaftlichen entwicklung*, Leipzig: Duncker & Humboldt. English translation, *The Theory of Economic Development*, Harvard, 1934; 8th edn, 1968, Cambridge, MA: Harvard University Press.

Schumpeter, J. (1942) *Capitalism, Socialism and Democracy*, Harper & Brothers, Harper Colophon edition 1975.

Science and Technology Policy Council of Finland (1996) *Finland: A Knowledge-Based Society*, Helsinki: Oy Edita Ab.

Science and Technology Policy Council of Finland (2000) *Review 2000: The Challenge of Knowledge and Know-How*, Helsinki: Oy Edita Ab.

Science and Technology Policy Council of Finland (2003) *Knowledge, Innovation and Internationalisation*, Helsinki: Oy Edita Ab.

Toivanen, H. (2000) 'Software Innovation in Finland', VTT, Group for Technology Studies. Working Papers No. 52/00, VTT, Espoo.

Törnudd, E. (1958) 'Teknillisestä tutkimuspanoksesta meillä ja muualla', *Paperi ja Puu*, No. 5, pp. 269–71.

Part IV
Recent Trends in Innovation

9
The International Dimension of Innovation Process: Evidence of Finnish Innovation Data

Nina Rilla

Introduction

Shortened product life cycles, rapid diffusion of new technology and increasing multidisciplinarity of new technology are shaping business and, at the same time, the innovation environment (Forrest, 1990; Narula, 2004). In order to respond to these challenges companies are often forced to look for external know-how and knowledge sources because of insufficient internal innovation resources (Howells, 1999; Tether, 2002). In order to survive companies need to open up their innovation process and to enter global innovation networks either through setting up their own research and development (R&D) unit, or forming R&D cooperative alliances with external partners (OECD, 2008). Even though it seems that companies' innovation processes have opened and become more international in recent years (OECD, 2008), innovation data used in this chapter challenges this picture.

R&D collaboration has traditionally been addressed from three main theoretical backgrounds: transaction costs, industrial organization theory and strategic management (Hagedoorn *et al.*, 2000). While transaction cost economics perspective tries to explain the reasons for firms to organize R&D internally and industrial organization theory to explain the resource allocation and economic welfare aspects of interfirm R&D collaboration, the strategic management approach concentrates more on analyzing the coordination and strategic positioning of R&D collaboration (Hagedoorn *et al.*, 2000). However, a strict division between these approaches should not be made since they are largely mixed and supported by each other. The strategic management perspective emphasizes, for example, the competitiveness, capability and networking side of research collaboration. Firms engage in R&D collaboration in order to broaden their own activities and resources, as well as maintain or enhance their competitive, often technological, strength (Hagedoorn *et al.*, 2000.) Innovation demands such complementary knowledge that is not usually economically feasible to develop all required know-how in-house (Teece, 1986), but, on the other hand, collaboration in innovation also requires ability to integrate and combine various types of knowledge inputs (Cohen and Levinthal, 1990).

The aim of this study is to explore the degree of collaboration in Finnish innovations in general, and the international dimension of innovation process in particular. This chapter places special attention on foreign collaboration and the extensiveness of foreign innovation networks, and analyzes the role of foreign collaboration in innovation process. Moreover, the impact of collaboration on innovation is touched upon. The main contribution of this study is that it places emphasis on innovation networks both domestic and foreign. International R&D collaboration is one form of openness and therefore an important issue from the policy and managerial perspective. This study will be informative for companies' management, improving its capacity to take advantage of external knowledge, both domestic and foreign, and innovation dynamics under various constellations, and also to decision-makers to enable them to produce more targeted policy incentives for companies active in innovation networks.

In this chapter the research partnerships are understood as collaboration with external partners during the innovation process. This study does not, however, take into account the research arrangements, such as strategic technical alliances as usually addressed in strategic management, but rather concentrates on the scope of R&D collaboration.

The study is organized in the following manner. The different collaboration arrangements and motives for collaboration are briefly outlined in the second section. The following section introduces the concept of innovation network, and the fourth section discusses the methodology and data. The fifth section introduces and discusses the results, and the last section offers conclusions and suggestions for future research.

Different collaboration forms

According to Hagedoorn's (1993) review of the motives and modes of technology partnering, monitoring of emerging technologies and general developments in science are among the main motivators in the forming of technology alliances and cooperating in innovation. Innovative companies simply need to stay in the forefront and have an access to sometimes-restricted diffusion of scientific and technological knowledge, which would not be available without the cooperative relationships with universities and research organizations. Second, also the inherent factors of uncertainty and risk, as well sharing costs, in R&D are among pushing factors in the forming of cooperation alliances (Mowery *et al.*, 1996). In general these are realized as concrete innovation projects, which might lead to shortened innovation processes and product life cycles. The third set of motives relates to the commercialization phase, for example creating new markets and products, and providing market entry (von Hippel, 1988; Mowery *et al.*, 1996; Hagedoorn *et al.*, 2000).

Formal and informal collaboration

Given that innovation often requires external expertise and development process is an interactive process, companies need to engage in collaboration and form various collaborative arrangements. Besides, know-how, whether knowledge or technology, is often available only on exchange basis (Veugelers, 1997). Hagedoorn (1993) suggests the cooperative alliances to develop and sustain technological competitiveness include university and research institute research agreements, collaborative R&D agreements and limited partnerships with private actors, that is, suppliers, customers or competitors. Moreover, inward technology licensing is a valuable means to strengthen innovation activities. On the other hand, collaboration may also be realized through client-sponsored research, for instance, with a larger competing company in order to acquire funding to develop and sustain core technology in-house (Forrest, 1990; Hagedoorn, 1993). Many of these collaboration forms are targeted at strengthening the technological expertise in a company's research, and are important collaboration forms, particularly to science-based companies. However, a firm's ability to evaluate and exploit the external knowledge is dependent on the level of prior related knowledge (Cohen and Levinthal, 1990).

Besides the above-introduced less equity-based or non-equity-based forms of innovation collaboration, in some cases the joint research is carried in the higher involvement intensive modes. Examples of such equity-based forms are formation of research joint venture or corporation, as well as acquisition or merger of research-intensive companies (Hagedoorn *et al.*, 2000). However, innovation collaboration is not concentrated only on innovation development, but also alliances in marketing and distribution, as well as manufacturing, are formed to complement a firm's own capabilities, for example, in the commercialization phase (Forrest, 1990; Jones, 2001). Moreover, the formation of joint ventures with R&D and marketing functions as well as outward licensing are valid choices in market access. In general, innovation-related activities, like R&D cooperation, might well be the first cross-border business activities companies engage in (Jones, 2001).

One special group of collaborators is customers, whose involvement in innovation designed in particular to industrial markets is significant (von Hippel, 1988). As von Hippel's study indicates, innovation development may originate from users' ideas, or their needs and problems may have great impact on development, such as prototype construction or product modifications. The tight user–manufacturer interaction is understandable since industrial innovations usually have a great importance on users' production processes or their competitive position (Biemans, 1992). Compared to research alliances with university, the user collaboration might be carried out in less formalized arrangements, and actually take place through personal relationships.

In addition to formal forms of innovation collaboration, also some more informal, or passive, activities tightly tied to innovation development can be recognized. Ronstadt and Kramer (1983) discuss scanning activities, which include, for instance, attendance at scientific/technical conferences, hosting of in-house seminars with leading international speakers or creating a company's technical advisory panel of outside experts. They have also identified some cooperation agreements, for example, individual consulting agreements, with universities and research institutions as belonging to this group. In addition, attendance to technology trade fairs in order to attract external cooperation is categorized as technology forecasting.

As the above discussion presents, the interorganizational cooperation takes place in several ways, whether requiring the setting up formal organizational modes, or with a contractual agreement basis, not to mention informal cooperative arrangements. In this study the innovation cooperation is understood as collaboration with external actors involved in a company's innovation process, not paying specific attention to the form.

Innovation networks

The innovation network consists of the cooperation arrangements with various actors, in which the position of the focal company is determined by the relations with other actors (Biemans, 1992). Depending on the type of collaboration partners and interaction taken, the collaboration can be divided into vertical and horizontal (Håkansson, 1987).

Cooperation between partners belonging to the same production chain, for instance between manufacturer, customer or supplier, is labelled as vertical cooperation. On the other hand, horizontal cooperation commonly relates either to cooperation with competitors or educational institutions. The horizontal cooperation strategy is relevant in situations, in which both parties can benefit, for instance, from reduced risks, faster development processes, or complementing know-how (Håkansson, 1987; Geroski, 1992).

According to Håkansson (1992), the industrial network consists of companies linked together by their production and utilization of either complementing or competitive products. The network contains three main dependent elements: actors, activities and resources. Actors can be, for example, individual people, a company or group of companies. Actors perform activities and control resources, whereas activities performed by actors link different resources to each other. Activities can be divided into transformation or transaction activities. On the other hand, the resources represent a vital condition for all activities and their value is determined by the activity. Resources consist of physical, financial and human assets (Biemans, 1992).

Innovation networks can be informal or formal (Ahuja, 2000; Powell and Grodal, 2006), depending, for example, on the research cooperation agreements. Personal relations and ties, such as membership in a trade association, among various informal research arrangements, form an informal network, often referred to as a social network, consisting of social capital (e.g., Granovetter, 2005; Mu *et al.*, 2008). These relationships are interesting as they might evolve into longer-lasting formal alliances (Powell and Grodal, 2006). Even though these informal relations are extremely difficult to analyze since they are difficult to quantify, some studies have been able to show the informal cooperation in innovation to be even more prevalent and important than formal cooperation (e.g., Bönte and Keilbach, 2005).

Besides the sociological aspect of innovation networks, the geographical extensiveness of innovation networks is a highly relevant aspect, especially from a technology policy point of view. The current debate concerns the significance of home base as the primary location for innovation, as innovation development seems to gain more and more international dimensions (Gertler *et al.*, 2000). On the one hand, it has been noted that innovation is realized in tight regional clusters, in which the shared common regional culture and proximity facilitates face-to-face communication and learning, especially in the transfer of tacit knowledge (Gertler *et al.*, 2000; Narula, 2003). In fact, the study of Patel and Pavitt (1991), for instance, states that large share of innovations are actually developed in a home location. On the other hand, we can also find arguments for the internationalization of innovation from the techno-globalism perspective, which emphasizes that cross-border technology and know-how transfer has increased since the barriers for diffusion have lowered (Getler *et al.*, 2000). It should be though noted that the industry differences in terms of innovation are significant (Pavitt, 1984), which means that, for example, certain industries are more international in terms of innovation development and diffusion than others. Given that firms are relying on external linkages in their innovation actions, the most suited linkages may not be found from the home; innovation collaboration networks also extend to foreign bases. Moreover, innovation networks are hardly constant, collaboration arrangements change over time (Powell and Grodal, 2006), while also domestic and foreign embeddedness is bound to alter. Nowadays, production of a commercially successful product may be largely dependent on a firm's ability to organize and manage its innovation networks (Geroski, 1992) in general, and to facilitate both local and global knowledge exchange and transfer in particular (Bathelt *et al.*, 2002).

The strategic importance of innovation networks relies on the fact that all the network actors are connected to each other either through direct or indirect relations, and are therefore affected by the individual relationships and the overall structure of network partners. Knowledge exchanged in one cooperative partnership is indirectly entered in another alliance (Ahuja, 2000; Verspagen and Duysters, 2004). The present study takes a less thorough view of innovation networks, since it assesses only the dyadic collaborative relationships between different actors in the innovation environment. Furthermore, the analysis of activities (e.g., collaboration arrangements) and resources (e.g., type of knowledge or technology) are excluded from this study, as well as the informal and indirect relations being unexplored. In this study, innovation networks are understood to consist of cooperation activities with various actors during innovation development.

Data and methodology

The analysis is based on Finnish Innovation data (SFINNO), which contains information of Finnish innovations commercialized by Finnish companies between 1945 and 2007. The database is continuously updated, and contains currently information of nearly 4500 innovations. The data is gathered using literature-based innovation output method

Table 9.1   Descriptive information of sample

	N	%
Innovations (total)	1074	
Firms (total)	767	
Collaboration (total)	920	85.7
of which domestic	880	95.6
of which foreign	648	70.4
Exported innovations	641	59.7

Table 9.2   Industry division of sample (note 1)

High-tech.	High-medium tech.	Medium-low tech.	Low tech.	KIS*	Other industries	Other services	N=
10.0%	26.4%	11.3%	13.2%	17.7%	7.6%	13.7%	1073

* KIS = knowledge intensive services

(LBIO), meaning systematic identification of innovations from selected industry-specific professional journals, expert opinions and companies' annual reports. After identification, the innovation developer (i.e., company) is approached with an invitation to fill in an online questionnaire. The definition given for innovation in this study follows OECD *Oslo Manual* (2005) definition, which states that innovation is an invention commercialized at market by a business or equivalent.

The analysis exploits the SFINNO survey data from 1985–2007, that is, those innovations for which a questionnaire was filled in and returned. The total size of sample is 1074 innovations. These innovations were developed and commercialized by 767 companies. All together, in 85.7 per cent of innovations the development has involved collaboration at least with one external partner. Most of the innovation development processes including collaboration have involved at last one domestic collaboration partner (95.6 per cent), while in 70.4 per cent of innovations collaboration with at least one foreign actor has taken part. Moreover, slightly over half of the innovations in the sample are exported. Table 9.1 presents descriptive information about the sample.

In order to get an impression of sectoral division of innovation collaboration, industries were divided according to their technological intensity (see Appendix 1 for detailed division of industries). From Table 9.2 we see that the largest share of innovations belong to high–medium technology

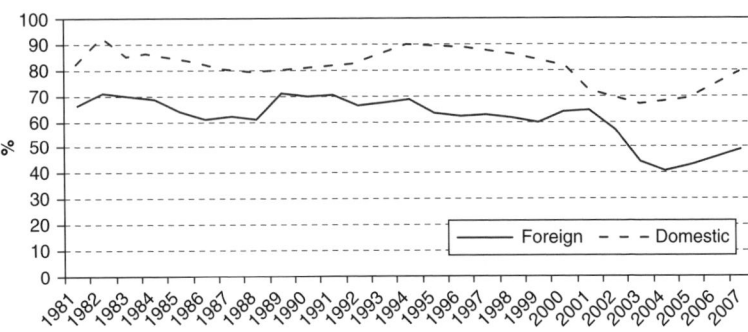

Figure 9.1   Evolution of domestic (880) and foreign (648) collaboration in Finnish innovations. Share of innovations including collaboration, presented in three years averages

sectors, which include, for example, manufacture of basic metals (SIC 24) and motor vehicles (SIC 29). The second largest group, the knowledge-intensive sectors, include sectors like telecommunications (SIC 61) and scientific research and development (SIC 72). In general, this division describes well the Finnish industrial structure.

Results and discussion

Longitudinal evolvement of innovation collaboration

To begin with, the data on collaboration in Finnish innovations indicates that innovations are developed rather in collaboration with domestic than foreign partners, see Figure 9.1. However, more interesting information is provided on a graph of the evolvement of innovation collaboration over the past twenty years. According to Finnish innovation data, collaboration in innovation declined in the beginning of the 21st century but has recovered again from 2005 onwards. Even though some recent studies report that R&D cooperation has become more common as a result of higher costs and technological complexity (e.g., Busom and Fernández-Ribas, 2008), innovations in Finland seem to involve less interorganizational cooperation than ten years ago. However, the deep decline shown between years 2000 and 2001 is partly due to the narrow innovation data on those years, although this does not fully explain the decline.

On the other hand, the collaboration among the different-sized companies has stayed fairly constant over the years, as Figure 9.2 reveals. It is

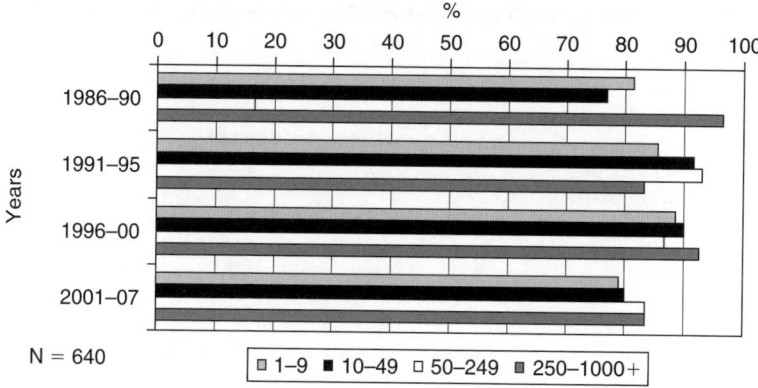

Figure 9.2 Share of innovations including collaboration divided by firm size (note 2)

interesting to notice that collaboration in innovation among the largest companies has lessened since the period 1986–90. Another important point to note is the modest collaboration activity among the largest small companies (50–249 employees) during the period 1986–90; however, collaboration climbed to the level of others in the subsequent period. The decline could be caused by the unavailable company data on those years. Among the smallest companies we could even see a slight increase in collaboration since the 1980s until the 2000s. Although collaboration among the micro companies (less than nine employees) and 10 to 49-employee firms is relatively strong, this information should be interpreted with caution. In general, the smallest companies represent the majority of Finnish company population; therefore also our data contain a high number of micro companies.

Even though there has been discussion recently about companies' increased concentration on core competencies and tightening of the value chain (Sako, 1994), this does not show in increased vertical cooperation. We would expect to detect an increase, especially in the large companies' innovation collaboration with supplier and subcontractors (Tether, 2002), but the opposite seems to have occurred. Vertical cooperation, in particular, has stayed rather constant over the years, whereas horizontal cooperation has fluctuated (see Appendix 2).

Innovation networks

As companies engage in collaboration with one partner, the likelihood of increasing the network relations grows (Fritsch and Lukas, 2001). Often

Table 9.3   Average size of innovation collaboration networks presented by firm size

	Firm size by no. of employees				
	1–9	10–49	50–249	250 >	n=
Average number of collaboration partners	3,3	3,1	3,6	3,0	754
Foreign partners Mode = 1	2,7	2,5	3,0	2,6	456
Domestic partners Mode = 3	3,6	3,7	4,2	3,5	616

Note: the division of innovations by firm size groups is based on innovations, which means that one company may appear several times if it has developed several innovations

innovation development requires expertise from several areas, as well as at different phases of development. In Table 9.3 we observe that the number of collaborative partners is quite constant among the different firm-size groups. The largest small companies (50–249 employees) have the most extensive innovation networks, while also micro companies (less than nine employees) have relatively extensive networks, including overseas. However, innovation networks are largely domestic, since they contain most commonly only one foreign collaboration partner, as compared to three domestic partners. Even though most innovations involve just one collaboration partner abroad, various innovations are still developed in networks containing several foreign partners as the high average network size implies.

As regards the extensiveness of innovation networks, it has been shown that the number of R&D cooperation parties has a positive effect on realizing product innovations, as well as that their combination enhances research productivity (Becker and Dietz, 2004).

Significance of research collaboration

Considering the complexities related to innovation cooperation, such as knowledge transfer, we are hardly able to assume collaboration always to be only beneficial on innovation. In order to assimilate new know-how and knowledge that the collaborative agreements are formed to create, some prior knowledge is essential (Lane and Lubatkin, 1998).

In general, the vertical collaboration, that is, customers, subcontractors, suppliers and own-concern companies, is seen as more significant than horizontal collaboration with universities, research institutes and competitors in innovation development (Table 9.4). Customer

Table 9.4 Share of innovations involving significant foreign and domestic collaboration divided by partner (note 3)

	Own concern	Customers	Subcontractors	Suppliers	Universities	Research institutes	Competitors	Consultants
Foreign	28%	48.7%	21%	42.9%	8.1%	11.2%	10%	6.6%
n=	75	544	533	77	532	529	538	527
Domestic	23.3%	67.1%	37.6%	35.8%	28.5%	17.7%	6%	15.5%
n=	583	608	599	134	512	603	586	595

collaboration is evidently the most important cooperation form in innovation. This applies to both domestic and foreign collaboration. In 67.1 per cent of innovations in which domestic collaboration has taken place, collaboration is rated significant, whereas among foreign collaboration slightly under half of cases (48.7 per cent) have considered such collaboration significant. If we look at foreign basic research collaboration (i.e., universities and research institutes), we note that collaboration with research institutes located abroad is seen as more important in innovation than foreign university cooperation. On the other hand, the opposite significance has been given to domestic horizontal collaboration, that is, university collaboration dominates. In contrast with significance, the non-significance of collaboration (e.g., with domestic competitors or foreign universities) raises concern. Either the outcomes of companies' collaborative agreements have been very unsatisfying, therefore trivial to innovation process, or there exists response bias in the data.

In spite of the dominance of domestic collaboration, cooperation with competitors located abroad seems to be relatively more important compared to domestic counterparts. Considering the novelty of innovations, this is understandable since most of the competitors are likely to locate abroad. In addition, intense competition in the home country might make cooperation with foreign actors more feasible (Lhuillery and Pfister, 2009). Overall, cooperative agreements with competitors seem to be fairly unimportant in collaboration portfolios, although engaging in collaboration with competitors might be highly beneficial to innovation as both parties' needs are close to each other. On the other hand, closeness of competence areas might also turn out to be one of the main reasons

Table 9.5   Share of significant foreign and domestic collaboration in innovations involving collaboration presented by firm size (note 4)

%	Firm size by no. of employees							
Foreign	**1–9**	***n=***	**10–49**	***n=***	**50–249**	***n=***	**250>**	***n=***
Own concern	18.5	*27*	31.3	*16*	25.0	*16*	43.8	*16*
Customers	44.7	*190*	55.4	*112*	60.0	*65*	44.6	*177*
Subcontractors	17.7	*186*	20.7	*111*	27.0	*63*	22.5	*173*
Suppliers	42.9	*28*	29.4	*17*	53.3	*15*	47.1	*17*
Universities	8.2	*184*	4.5	*111*	9.2	*65*	9.9	*172*
Research institutes	11.4	*184*	7.3	*109*	7.8	*64*	14.5	*172*
Competing companies	8.9	*190*	11.7	*111*	11.9	*67*	8.2	*170*
Consultants	7.6	*184*	8.3	*109*	4.7	*64*	5.3	*170*
Domestic	**1–9**	***n=***	**10–49**	***n=***	**50–249**	***n=***	**250>**	***n=***
Own concern	18.0	*206*	16.0	*125*	27.1	*70*	33.0	*182*
Customers	66.7	*222*	73.1	*130*	72.2	*72*	61.4	*184*
Subcontractors	40.2	*219*	38.3	*128*	40.0	*70*	33.0	*182*
Suppliers	29.6	*54*	47.1	*34*	36.4	*22*	33.3	*24*
Universities	33.8	*154*	27.6	*134*	23.4	*107*	27.4	*117*
Research institutes	21.7	*217*	13.6	*132*	20.0	*70*	15.2	*184*
Competing companies	6.6	*212*	3.9	*129*	8.7	*69*	5.7	*176*
Consultants	18.1	*215*	16.0	*131*	17.4	*69*	11.1	*180*

for cooperation failure, as recent study by Lhuillery and Pfister (2009) reveals.

Even though foreign collaboration is less valued than domestic, we observe differences among the company groups of significant foreign collaborators. Table 9.5 illustrates significant cooperation arrangements for innovation involving collaboration among different firm-size classes. Companies of different size do collaborate with foreign partners fairly evenly, the vertical collaborative arrangements being the most relevant, whereas the importance of horizontal collaboration is less valued in innovation. For micro companies the customer collaboration is clearly the most significant collaborative form among foreign partners, whereas horizontal collaboration seems to be less important in innovation. Although collaboration with foreign research institutes is clearly seen as more important than any other form of horizontal collaboration, only collaboration with own-concern firms located abroad is more relevant to large firms than micro companies, although the number of observations is rather low. The horizontal innovation collaboration is carried out largely domestically in all firms, except collaboration with competitors. The smallest companies engaged in domestic horizontal cooperation fairly intensively. In contrast to Veugelers and Cassiman's (2005) evidence on large companies' higher likelihood to engage in university collaboration, the results of Finnish data emphasize that also micro companies value university collaboration, in particular with domestic universities. On the other hand, the nature of collaboration is likely to vary between firms. Large companies have established R&D departments, whereas SMEs operate with smaller resources; therefore SMEs are more dependent on external expertise than large companies, whose main intention of horizontal collaboration might be to keep up with forefront research.

Vertical collaboration, which includes customers,subcontractors, supplier and firms belonging to same concern, is the prevailing cooperation arrangement in all size classes. In general, companies' vertical innovation networks seem more extensive geographically compared to horizontal collaboration networks. Consequently, as innovations are aimed at international markets, the value chain operations, such as subcontractor networks, are built on international basis as well. Horizontal cooperation is still largely domestic, possibly as a result of a need for tighter collaboration. In several cases collaboration requires face-to-face contacts and intense communication, which might be more fruitful with partners, for example, sharing the same cultural context. Second, access to universities located nearby, and also building informal networks as well as maintaining personal relations, is somewhat more straightforward as compared to doing so with institutions abroad. On the other hand, the accessibility of foreign research institutes could be lower due to their role in applied research, in which the collaboration arrangements are, perhaps, shorter and easier to handle compared to basic research performed in universities.

After all, in the discussion of innovation networks, it has been shown that R&D spillovers, that is, knowledge flows caused by cooperation, and proximity of R&D operations, have great impact in companies' R&D location choices (e.g., Feinberg and Gupta, 2004; Cassiman and Veugelers, 2002), and might partly explain the higher dominance of interorganizational cooperation with domestic parties. Moreover, the results show that collaboration with domestic customers is quite relevant and common, but collaboration with foreign customers is less significant. Given that many innovations are developed for international

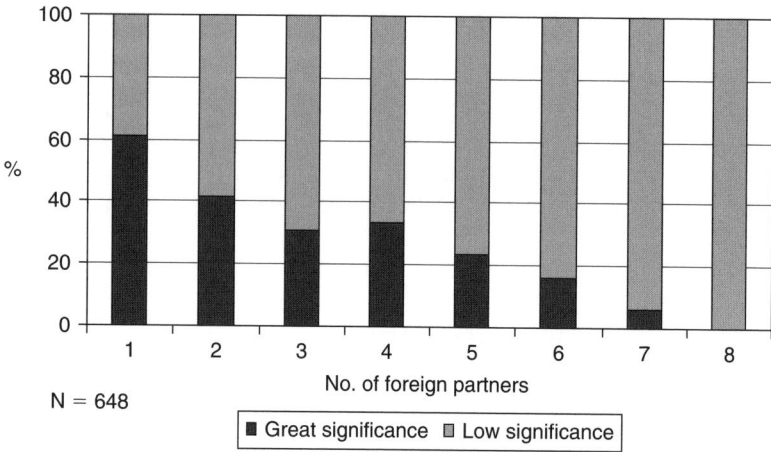

Figure 9.3   Significance of collaboration partners in different sized foreign networks (note 5)

markets, it is surprising to notice that proximity matters also in customer cooperation.

Overall, several studies emphasize large companies' eagerness to collaborate as being higher compared to smaller counterparts (Rothwell and Dodgson, 1991; Tether, 2002; Cassiman and Veugelers, 2002), but we do recognize the high importance of cooperative arrangements to the smallest companies as well. Because of the restricted innovation resources, micro companies are eager to form cooperative arrangements, but at the same time they are able to manage relatively extensive innovation networks considering their limited resources. The innovation networks of the smallest companies have high diversity; networks consist of both vertical and horizontal cooperation arrangements.

Size and significance of collaborative networks

Companies seek various kind of know-how from external sources in order to succeed in innovation; however, organizing and managing extensive innovation networks is challenging. As Figures 9.3 and 9.4 illustrate, the significance of collaboration decreases according to the size of network, the more collaboration partners, the less important collaboration has been on innovation. A similar trend is seen in both foreign and domestic networks.

Given the tacit side of innovation collaboration, the decrease in importance might be explained by a couple of reasons: 1) construction and

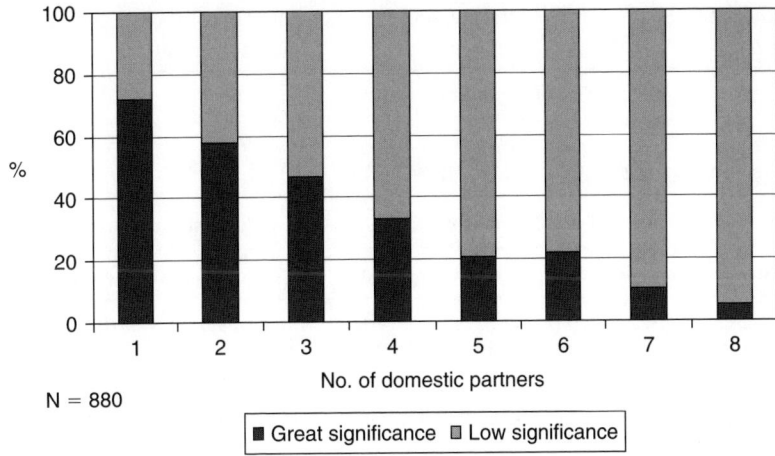

Figure 9.4 Significance of collaboration partners in different sized domestic networks (note 6)

maintaining of wide networks require resources; 2) knowledge required in innovation is not widely spread. Creating a trustworthy and well-working collaborative relationship is a long but also demanding process (Häusler *et al.*, 1994). Besides, the innovation collaboration is delicate, and the required knowledge is not always easily available and transferable. In general, innovation collaboration includes high levels of risk and uncertainty, mainly because of the quality of knowledge exchanged in research projects is mostly unknown beforehand (Howells *et al.*, 2008). When companies have succeeded in building one or a few valuable cooperation relationships, they rather develop and cherish those than spread their scarce resources to maintain several less vital relations. Furthermore, the advantage of proximity diminishes as amount of innovation partners in network increases, as we observe a decrease in collaboration significance also in domestic networks.

Impact of collaboration on innovation

As can be concluded from the above discussion, innovations still require tight interorganizational cooperation with certain partners, regardless of the diminishing trend in collaboration over the past twenty years. In order to have a deeper understanding of the impact of collaboration on innovation, innovations have been divided into three classes according

Table 9.6   Share of different innovation types according to collaboration network size

Size of innovation network

	n=	None	Low	Heavy	%			
All innovations	1074	14.3	42.1	43.6	100			

	Foreign network				Domestic network			
Innovation type	n=	None	Low	Heavy	%	None	Low	Heavy	%
Science origin	209	34.0	25.8	40.2	100	13.9	15.8	70.3	100
Market origin	836	36.1	34.7	29.2	100	16.4	21.8	61.8	100
New to firm	667	40.3	33.4	26.2	100	16.5	25.6	57.9	100
Global novelty	757	33.6	37.3	29.2	100	15.5	24.6	60.0	100

to the extensiveness of collaboration network: on-collaborators, low collaborators and heavy – collaborators (note 7).

It seems that innovations with global-market novelty demand wide domestic collaboration but less foreign cooperation than expected (Table 9.6). In addition, foreign collaboration in innovations new to the global market is more common than in new-to-firm innovations, also with domestic partners. Looking at other types of innovations, we see that science-originated innovations involve rather extensive foreign cooperative relations (40.2 per cent) compared to market originating innovations, in which only 29.2 per cent of innovations have involved more than three foreign partners. These results support the findings of Tether (2002), who states that cooperation in new to global market innovations is more common than in innovations only new to firm. Also Freel (2003) has reported similar finding; share of foreign collaboration is likely to increase according to innovation novelty. The new-to-firm innovations are often modifications and improvements to existing innovations, which are produced with internal resources; instead, innovations with higher novelty require more complex but at the same time less rarely available know-how. However, this external knowledge tends to be acquired from domestic rather than foreign sources. It could, though, be that the innovation developer's strategy is to look for partners first from institutions close by for the advantages proximity provides, and if not successful then move on to establishing cooperation agreements with actors at a longer distance.

Table 9.7 Different-sized collaboration networks divided by innovation's development time from idea to commercialization

Network size	Development time				
	Same year %	1–2 years %	3–5 years %	6–9 years %	10+ years %
n=	106	422	297	121	61
Foreign None	59.4	41.9	30.0	36.4	21.3
Low	18.9	34.6	43.4	34.7	29.5
Heavy	21.7	23.5	26.6	28.9	49.2
	100%	100%	100%	100%	100%
Domestic None	39.6	18.2	11.8	11.6	6.6
Low	11.3	28.0	28.3	22.3	21.3
Heavy	49.1	53.8	59.9	66.1	72.1
	100%	100%	100%	100%	100%

It has been suggested that cooperation affects the innovation development time positively, that is, it shortens the process from idea to market commercialization (e.g., von Hippel, 1988; Hagedoorn and Schakenraad, 1990). This finding is supported by our data as well to a certain extent. The major share of innovations are developed in one to two years (41.9 per cent of the total number of innovations), regardless of the geographical scope or number of cooperation partners (Table 9.7). On the other hand, innovations which have extremely short development time, that is, less than one year, involve less collaboration compared to innovations with longer development times. The collaboration relations, especially the formal engagements, take time and effort to arrange. Therefore, we could assume to see a lesser amount of collaboration in innovations, which have an extremely short, and at the same time straightforward, development process. Optimization of the size of collaboration network seems to expedite development time; however, we should remember that development processes vary between innovation types as well as industries. Therefore generalizations regarding length of innovation process should be made with caution.

Impact of collaboration on internationalization

The role of networks in companies' internationalization is widely recognized in the international business literature (see, e.g., Johanson and

Table 9.8   Different sized collaboration networks divided by innovation's time
from commercialization to exporting

Network size	Time for exporting			
	Same year %	1–2 years %	3–5 years %	5+ years %
n=	285	261	73	22
Foreign None	24.6	29.5	30.1	18.2
Low	44.9	37.2	45.2	45.5
Heavy	30.5	33.3	24.7	36.4
	100%	100%	100%	100%
Domestic None	14.4	11.5	13.7	4.5
Low	29.1	24.1	26.0	13.6
Heavy	56.5	64.4	60.3	81.8
	100%	100%	100%	100%

Mattson, 1988; Coviello and Munro, 1997). Industrial markets are seen
as networks that consist of formal and informal interlinked relations
between companies. Some of the relationships might assist companies
to enter foreign markets, providing opportunities and motivation to
internationalize, but, at the same time, also inhibit international development (Coviello and Munro, 1997). As regards collaborative innovation
networks, we could argue that cooperative relations in the early stages of
merchandise development and production are valuable components of
a company's operational environment, that is, its network (Håkansson,
1992).

In Table 9.8, an innovation's time to export has been divided according to size of domestic and foreign networks measured by cooperation
partners. This reveals that exporting is likely to begin soon after innovation commercialization, as in a major share of innovations (85 per
cent) exports start during the same year or in one to two years from
commercialization. Most of the rapidly exported innovations, that is,
where exporting starts more or less in conjunction with commercialization, have engaged in collaboration with one to two foreign partners
and undertaken more extensive domestic cooperation during the development. In contrast to duration of development process, engaging in
collaboration seems to be beneficial to start of exports.

In addition to number of foreign collaboration partners also the quality of collaboration seems to matter in the start of exporting. Figure 9.5
reveals that significance of foreign collaboration is the highest among
innovations, in which exporting takes place more or less concurrently

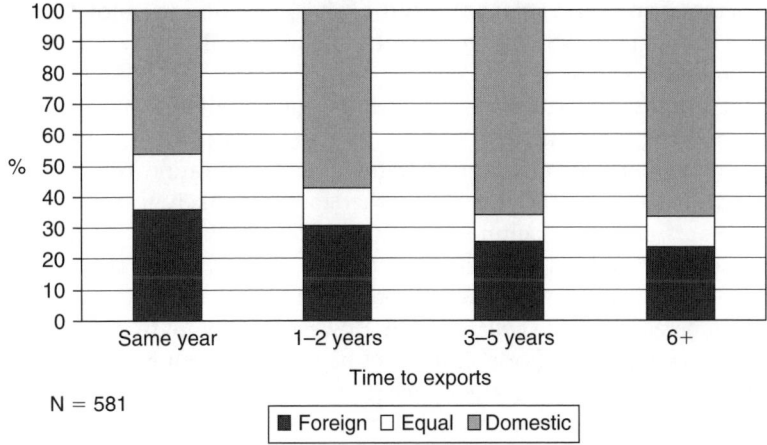

Figure 9.5 Significance of different collaboration types in innovation on duration from commercialization to exports

with commercialization. The importance of cooperation with partners abroad diminishes, while the importance of domestic collaboration increases, the longer time the start of exports takes.

These results imply that cooperation with foreign actors during the innovation process creates valuable connections and opportunities that companies need when starting export operations. For example, close relations to customers during the innovation development aid and fasten the start of exports as demand is created simultaneously with innovation cooperation. To conclude, foreign collaboration arrangements enable smooth exporting but seem to complicate development by lengthening the innovation process.

Concluding comments

The results prove that innovation development requires still strong inter-organizational collaboration; however, the nature of cooperation has changed towards being more strategic. The expectations of the outcomes and importance of cooperative arrangements have increased; therefore, it is natural that companies put emphasis on partner evaluation and selection. Compared to several other European countries, innovation development in Finland involves a high share of interorganizational cooperation, which could be explained by a strong 'national culture of cooperation', which is supported by, for example, R&D collaboration schemes (van Beers *et al.*, 2007). Although cooperation in innovation is seen as important, knowledge and information are still largely developed in domestic networks.

As regards opportunities provided by globalization, this has not affected the most efficient ways of sourcing international knowledge in innovation. Collaboration in general, and innovation cooperation in particular, is a delicate activity; therefore, it is likely to be affected by factors related to culture or personal relations in favour of domestic rather than foreign interorganizational collaboration. This makes domestic collaborative relations more vital to innovation development as compared to foreign collaboration. However, this is not to suggest that foreign cooperative efforts should not be pursued because of their lessened importance – quite the opposite. The strategic aims could just be different to what is expected from domestic collaboration. Considering innovation collaboration from the network perspective, the significance of foreign collaboration is realized, for instance, through the easier access and establishment in foreign markets initiated by collaborative contacts.

We observed that especially the small Finnish innovative companies engaged in cooperation, even quite extensively. Therefore, also innovation policy should focus on supporting and, first of all, providing the means for them to consummate meaningful innovation collaboration relations with foreign as well as domestic actors. On the other hand, larger companies might have more efficient channels and means for knowledge transfer between partners, which enables the benefits of foreign knowledge sourcing to be distributed wider in the local innovation environment, including to micro companies, for instance, through subcontracting arrangements.

Regardless of the considerable importance of collaboration in innovation development and all the current open innovation discussion (e.g., Chesbrough *et al.*, 2006), cooperation is not by any means a compulsory condition for innovation (Freel and Harrison, 2006). There are several successful innovations which have not engaged in any collaborative activities in any phases of the innovation development, as was also demonstrated in this study. Given that innovations consist of companies' core competencies and provide their competitive advantage, they are likely to be developed under a veil of secrecy, suggesting that collaboration is carried out for various reasons, of which keeping track of competitors' operations is not surely the least important motive.

To conclude, innovation collaboration strategies in companies, as well as policy incentives targeted to enhance cooperation in innovation development, should emphasize quality rather than quantity.

Future research

The current simple and descriptive analysis included only formal collaboration with certain actors, which gives a partial picture of collaborative arrangements in innovation. In order to have a complete understanding of the extent of collaboration and innovation networks, forms of informal R&D collaboration should be included in the investigation. However, a more qualitative research approach is needed to carry this out. Similarly, information about and understanding of foreign collaboration patterns in innovation would need more detailed and informative data. For example, data on collaboration forms, geographical scope and duration would allow us to arrive at a more precise conclusion of international sourcing of knowledge. This would give the possibility of investigating the international collaboration strategies of companies more thoroughly, and the interconnectedness of internationalization and innovation activities in general. This is especially relevant to SMEs, whose international knowledge-sourcing strategies are largely unexplored.

Appendix 1: Industry classification (based on SIC 2002)

1 = High-technology industries (SIC 30, 32, 33)
2 = High–medium technology industries (SIC 24, 29, 31, 34–5)
3 = Medium–low technology industries (SIC 23, 25–8)
4 = Low-technology industries (SIC 15–22, 36–7)
5 = Knowledge-intensive services (SIC 61–2, 64, 65–7, 72, 74.2, 74.3)
6 = Other industries (SIC 10–14, 45, 60, 70, 73, 80, 85)
7 = Other services (SIC 40–1, 50–2, 63, 71, 74–5, 90, 92)

Sources: Statistics Finland classification (Eurostat, e.g., Statistics in Focus 13/2006); OECD

Appendix 2

Change in vertical and horizontal collaboration types over time. Vertical collaboration: own concern firms, customers, subcontractors and suppliers. Horizontal collaboration: universities, research institutes, competitors and consultants.

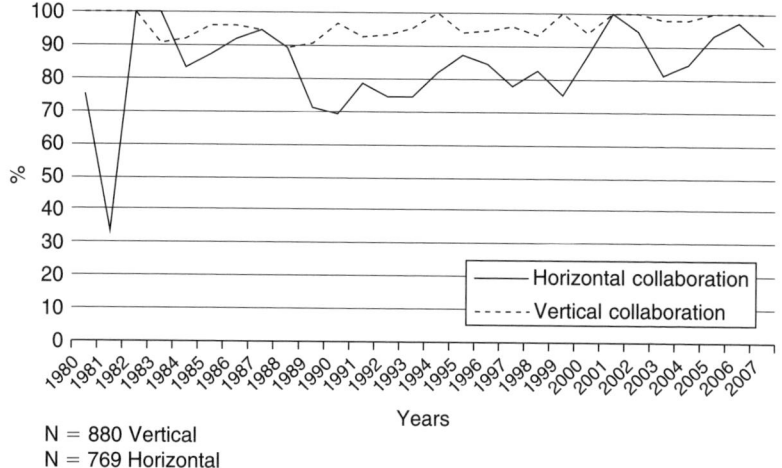

N = 880 Vertical
N = 769 Horizontal

Appendix 2   Share of vertical and horizontal collaboration of all innovations involving collaboration

Notes

1. Division is made according to SIC class of company responsible for innovation development and commercialization.
2. Firm size is measured in number of employees at the time of innovation commercialization.
3. Significant innovations present innovations in which collaboration has been rated as significant or very significant (scale from not significant to very significant) for innovation development.
4. Significant innovations present innovations in which collaboration has been rated as significant or very significant (scale from not significant to very significant) for innovation development. Comparison utilized sample of innovations, of which the firm size class was available.
5. In great significance average is between 2 and 3; in low significance, average is 0–1.
6. In great significance average is between 2 and 3; in low significance, average is 0–1
7. The size of collaboration network is measured by the number of collaboration partners; the group 'none' do not have any innovation collaboration, low

collaboration networks refer to one to two partners and heavy networks consist of more than three collaboration partners.

References

Ahuja, Gautam (2000) 'Collaboration Networks, Structural Holes, and Innovation: A Longitudinal Study', *Administrative Science Quarterly*, Vol. 45, No. 3, pp. 425–55.

Bathelt, Harald, Malmberg, Anders and Maskell, Peter (2002) 'Clusters and Knowledge: Local Buzz, Global Pipelines and the Process of Knowledge Creation', DRUID Working Paper No. 02-12, Copenhagen.

Becker, Wolfgang and Dietz, Jürgen (2004) 'R&D Cooperation and Innovation Activities of Firms: Evidence for the German Manufacturing Industry', *Research Policy*, Vol. 33, No. 2, pp. 209–23.

Biemans, Wim G. (1992) *Managing Innovation within Networks*, London: Routledge.

Bönte, Werner and Keilbach, Max (2005) 'Concubinage or Marriage? Informal and Formal Cooperations for Innovation', *International Journal of Industrial Organization*, Vol. 23, pp. 279–302.

Busom, Isabel and Fernández-Ribas, Andrea (2008) 'The Impact of Firm Participation in R&D Programmes on R&D Partnerships', *Research Policy*, Vol. 37, pp. 240–57.

Cassiman, Bruno and Veugelers, Reinhilde (2002) 'R&D Cooperation and Spillovers: Some Empirical Evidence from Belgium', *The American Economic Review*, Vol. 92, No. 4, pp. 1169–84.

Chesbrough, Henry, Vanhaverbeke, Wim and West, Joel (2006) *Open Innovation: Researching a New Paradigm*, Oxford: Oxford University Press.

Cohen, W. M. and Levinthal, D. A. (1990) 'Absorptive Capacity: A New Perspective on Learning and Innovation', *Administrative Science Quarterly*, Vol. 35, No. 1, pp. 128–52.

Coviello, Nicole and Munro, Hugh (1997) 'Network Relationships and the Internationalisation Process of Small Software Firms', *International Business Review*, Vol. 6, No. 4, pp. 361–86.

Feinberg, Susan E. and Gupta, Anil K. (2004) 'Knowledge Spillovers and the Assignment of R&D Responsibilities to Foreign Subsidiaries', *Strategic Management Journal*, Vol. 25, No. 8/9, pp. 823–45.

Forrest, Janet E. (1990) 'Strategic Alliances and the Small Technology-Based Firm', *Journal of Small Business Management*, Vol. 28, No. 3, pp. 37–45.

Freel, Mark (2003) 'Sectoral Patterns of Small Firm Innovation, Networking and Proximity', *Research Policy*, Vol. 32, pp. 751–70.

Freel, Mark and Harrison, Richard (2006) 'Innovation and Cooperation in the Small Firm Sector: Evidence from "Northern Britain"', *Regional Studies*, Vol. 40, No. 4, pp. 289–305.

Fritsch, Michael and Lukas, Rolf (2001) 'Who Cooperates on R&D?', *Research Policy*, Vol. 30, No. 2, pp. 297–312.

Geroski, P. A. (1992) 'Vertical Relations Between Firms and Industrial Policy', *The Economic Journal*, Vol. 102, No. 410, pp. 138–47.

Gertler, M., Wolfe, D. and Garkut, D. (2000) 'No Place Like Home? The Embeddedness of Innovation in a Regional Economy', *Review of International Political Economy*, Vol. 7, No. 4, pp. 688–718.

Granovetter, Mark (2005) 'The Impact of Social Structure on Economic Outcomes', *The Journal of Economic Perspectives*, Vol. 19, No. 1, pp. 33–50.

Hagedoorn, J. (1993) 'Understanding the Rationale of Strategic Technology Partnering: Interorganizational Modes of Cooperation and Sectoral Differences', *Strategic Management Journal*, Vol. 14, No.5, pp. 371–85.

Hagedoorn, John and Schakenraad, Jos (1990) 'Inter-firm Partnerships and Cooperative Strategies, in C. Freeman and L. Soete (eds), *New Explorations in the Economics of Technical Change*, London: Pinter, pp. 3–37.

Hagedoorn, John, Link, Albert N. and Vonortas, Nicholas S. (2000) 'Research Partnerships', *Research Policy*, Vol. 29, pp. 567–86.

Håkansson, Håkan (1987) *Industrial Technological Development: A Network Approach*, London: Croom Helm.

Håkansson, Håkan (1992) *Corporate Technological Behaviour: Cooperation and Networks*, London: Routledge.

Häusler, Jürgen, Hohn, Hans-Willy and Lütz, Susanne (1994) 'Contingencies of Innovative Networks: A Case Study of Successful Interfirm R&D Collaboration', *Research Policy*, Vol. 23, pp. 47–66.

Howells, Jeremy (1999) 'Research and Technology Outsourcing', *Technology Analysis & Strategic Management*, Vol. 11, No. 1, pp. 17–29.

Howells, Jeremy, Gagliardi, Dimitri and Malik, Khaleel (2008) 'The Growth and Management of R&D Outsourcing: Evidence from UK Pharmaceuticals', *R&D Management*, Vol. 38, No. 2, pp. 205–19.

Johanson, Jan and Mattsson, Lars-Gunnar (1988) 'Internationalisation in Industrial Systems: a Network Approach', in N. Hood and J.-E. Vahlne (eds), *Strategies for Global Competition*, London: Croom Helm, pp. 298–314.

Jones, Marian V. (2001) 'First Steps in Internationalisation: Concepts and Evidence from a Sample of Small High-technology Firms', *Journal of International Management*, Vol. 7, No. 3, pp. 191–210.

Lane, Peter and Lubatkin, Michael (1998) 'Relative Absorptive Capacity and Interorganizational Learning', *Strategic Management Journal*, Vol. 19, No. 5, pp. 461–77.

Lhuillery, Stéphane and Pfister, Etienne (2009) 'R&D Cooperation and Failures in Innovation Projects: Empirical Evidence from French CIS Data', *Research Policy*, Vol. 38, No.1, pp. 45–57.

Mowery, David C., Oxley, Joanne E. and Silverman, Brian S. (1996) 'Strategic Alliances and Interfirm Knowledge Transfer', *Strategic Management Journal*, Vol. 17, pp. 77–91.

Mu, Jifeng, Peng, Gang and Love, Edwin (2008) 'Interfirm Networks, Social Capital, and Knowledge Flow', *Journal of Knowledge Management*, Vol.12, No. 4, pp. 86–100.

Narula, Rajneesh (2003) *Globalization and Technology: Interdependence, Innovation Systems and Industrial Policy*, Cambridge: Polity Press.

Narula, Rajneesh (2004) 'R&D Collaboration by SMEs: New Opportunities and Limitations in the Face of Globalisation', *Technovation*, Vol. 24, No. 2, pp. 153–61.

OECD (2005) *Oslo Manual: Guidelines for Collecting and Interpreting Innovation Data* (3rd edn), Paris: OECD and Eurostat.

OECD (2008) *Open Innovation in Global Networks*, Paris: OECD.

Patel, Pari and Pavitt, Keith (1991) 'Large Firms in the Production of the World's Technology: An Important Case of "Non-Globalisation"', *Journal of International Business Studies*, Vol. 22, No.1, pp. 1–21.

Pavitt, K. (1984) 'Sectoral Patterns of Technical Change: Towards a Taxonomy and a Theory', *Research Policy*, Vol. 13, No. 6, pp. 343–73.

Powell, Walter W. and Grodal, Stine (2006) 'Networks of Innovators', in Jan Fagerberg, David Mowery and Richard Nelson (eds), *The Oxford Handbook of Innovation*, Oxford: Oxford University Press, pp. 57–85.

Ronstadt, Robert and Kramer, Robert J. (1983) 'Internationalizing Industrial Innovation', *Journal of Business Strategy*, Vol.3, No. 3, pp. 3–15.

Rothwell, Roy and Dodgson, Mark (1991) 'External Linkages and Innovation in Small and Medium-Sized Enterprises', *R&D Management*, Vol. 21, No. 2, pp. 125–38.

Sako, Mari (1994) 'Supplier Relationships and Innovation', in M. Dodgson and R. Rothwell (eds), *The Handbook of Industrial Innovation*, Cheltenham, Brookfield: Edward Elgar, pp. 268–74.

Teece, David J. (1986) 'Profiting from Technological Innovation', *Research Policy*, Vol. 15, No. 6, pp. 21–45.

Tether, Bruce (2002) 'Who Cooperates for Innovation, and Why: An Empirical Analysis', *Research Policy*, Vol. 31, No. 6, pp. 947–67.coop

Verspagen, Bart and Duysters, Geert (2004) 'The Small Worlds of Strategic Technology Alliances', *Technovation*, Vol. 24, pp. 563–71.

Veugelers, Reinhilde (1997) 'Internal R&D Expenditures and External Technology Sourcing', *Research Policy*, Vol. 26, pp. 303–35.

Veugelers, Reinhilde and Cassiman, Bruno (2005) 'R&D Cooperation Between Firms and Universities: Some Empirical Evidence from Belgian Manufacturing', *International Journal of Industrial Organization*, Vol. 23, No. 5–6, pp. 355–79.

van Beers, Cees, Berghäll, Elina and Poot, Tom (2007) 'R&D Internationalization, R&D Collaboration and Public Knowledge Institutions in Small Economies: Evidence from Finland and the Netherlands', DRUID Working Paper No. 07-12.

von Hippel, E. (1988) *Sources of Innovation*, New York: Oxford University Press.

10
When the Going Gets Tough: Failure of Innovative Businesses

Pekka Pesonen and Robert van der Have

Introduction

Innovation is widely believed to play a key role for the survival and competitiveness of firms. The significance of innovation has been stressed increasingly among academics in recent decades, and policy-makers and practitioners have rather pervasively adopted the view. Nowadays, continuous renewal by firms is considered as an essential organizational process in coping with technological progress (Teece *et al.*, 1997) and the positive effect of innovation as a cornerstone of renewal seems almost unquestionably accepted. Firms can pursue product innovations to go into new industries or markets, or introduce new technologies or product features in their existing domains to extract greater rents or obtain an advantage vis-à-vis their competitors (Burgelman and Sayles, 1986). Greve and Taylor (2000) mention innovation as a 'competitive weapon' to get new resources and competences when radically new product introductions undermine the incumbent technological regime.

If innovation is indeed at the very core of the success of firms, one may still question two things. First, if innovation is a necessary condition, is it also a sufficient condition for business performance? It is obvious that the answer to this first question is negative; empirical evidence shows, indeed, that innovation does not invariably lead to success and growth (see Barnett and Freeman, 2001). In fact, some firms vanish from the markets within only a few years or even during the same year they introduce an innovation. Second, if innovation lies at the hart of wealth-creation but it does not prevent firms from failing in this *per se*, can we then find certain aspects of innovation which may be related to such failure to extract value? On the answer to this second question, only very little is known at present. Cefis and Marsili (2006) conclude as well that it is worth looking in more detail at the characteristics of the innovation process and the nature of innovations in relation to the risk of failure after studying the effect of innovation on the survival of firms. In our chapter, we make an attempt to explore the issue of how the characteristics of single innovations, their individual innovation processes, and the firms that commercialize them can be related to business failures of innovative firms. Existing empirical work relating innovation to success or failure has thus far been rather fragmented in a conceptual sense. This chapter aims to contribute, first, by distinguishing the levels of a) the single product innovation, b) its distinct innovation process and c) the innovating firm. Second, we contribute to the empirical literature by applying a more precise definition of 'business failure' than is usual in related work. A third aspired contribution to this research topic is to study a group of significantly innovating firms, as opposed to applying the usual innovator–non-innovator distinction or relying on self-reported 'innovators', which may lack significant innovativeness from a more objective standpoint.

In the first section of our chapter, we discuss the most relevant theoretical and empirical literature for our study. As our study is quite explorative in nature, we will not derive any hypotheses here, but rather try to clarify the research issue at hand. In the second section we describe our data in more detail, including a descriptive analysis. The following section entails a description of our statistical model. In the fourth section, we look at the results of the statistical analysis. The final section concludes.

Theoretical and empirical background

Innovation and success

A myriad of research has linked innovation and success with each other in different ways in different lines of research. On the process and product level, the focus has been on finding the keys to a successful innovation.[1] The first study of this kind was project SAPPHO, which was conducted in the early 1970s. In the first phase of the study, comparing 29 pairs of successful and unsuccessful innovations, it was found that five main characteristics separate success from failure. These were innovator's comprehension of customer's needs; efficiency of development and communication; managerial capabilities; marketing capabilities and sales efforts (Freeman *et al.*, 1972). In the second phase of the SAPPHO project, the results were supported by a study extension of 43 project pairs (Rothwell *et al.*, 1974). Since SAPPHO, several studies have contributed in the field of success of innovation and analyzed different success factors. Studying 103 firms, Cooper (1979) identified important factors affecting new product success. Griffin and Page (1996), in turn, provided recommended measures for assessing whether an innovation is a success or a failure from the firm's point of view. They noted that there are differences between firms, depending on their strategies, in what is regarded as a successful innovation project. Process success factors were also used by Cooper and Kleinschmidt (1995), who found that the best firm-level performers apply certain type of practices, methods and structures on the innovation-process level. The success factors of innovation projects have been noted to differentiate according to characteristics of innovation, for instance, its novelty (see de Brantani, 2001; McDermott and O'Connor, 2002). In addition, there have been studies exploring attributes of successful innovation in different phases of the innovation process, for example, in idea generation (see, e.g., Goldenberg *et al.*, 2001) and the sources of innovation (see, e.g., Palmberg, 2006).

Much of the motivation behind 'new product development' studies has been to help firms concentrate on crucial elements for successful new processes/products that will in turn increase the probability of business success. Thus, this type of studies most often focuses on describing what to do (in order to achieve success) rather than describing what *not* to do (that can lead to failure). Where addressed, the failure factors have been mostly considered as being diametrically opposed to success factors, but typically in previous research the failure factors have not been the main interest.

On the firm level, innovation and its characteristics have been the subjects of examination in terms of explaining the success of businesses. Innovativeness has been found to have a positive effect on the profitability of the innovating firm (Geroski *et al.*, 1993), signifying that producing new products is in fact supporting firm success. There are also substantial differences in the longer-term profitability of innovating and non-innovating firms, in favour of the former. Profiting from innovativeness further in time from the actual introduction of innovation implies that the roots of profitability are deeper in firm's capabilities (see Teece *et al.*, 1997). Through the process of innovation and renewal, firms build new and enhance existing skills and know-how, which leads to longer-term benefits in addition to direct rents from selling a new product/service. Saarinen and Niininen (2000) studied the profitability of innovative firms as well, and found that it is influenced by different factors of the innovation process. Their main results suggest that a shorter development time of innovation and the use of new technologies in innovation are related to higher profits.

Furthermore, in studies dealing with innovation and firm survival, it has been noted that innovation has a positive effect on the probability of a firm's survival (Cefis and Marsili, 2006). This 'innovation premium', as called by Cefis and Marsili (2006), was noted to be larger for small firms, as well as young firms. This suggest that these types of firms seem to benefit most – in terms of survival – from developing unique offerings to markets. In addition, the growth rate and nature of technology of firms have been recognized as playing a major role in firm survival (Cefis and Marsili, 2005). On the level of individual innovations, research aiming to reveal factors of firm survival has been more limited in terms of number of studies. However, in the case of type of innovation, there is some evidence that firms commercializing a process innovation are more likely to survive than firms with product innovations (Cefis and Marsili, 2005; Huergo and Jaumandreu, 2004). This implies that firms developing their processes get a certain advantage through cost reduction or more robust capabilities which in turn enables them to stay competitive in the following years of innovation. Nevertheless, in studies of innovation and survival, failure has been typically included only in terms of firm exit (see, e.g., Cefis and Marsili, 2005). There has typically not been a distinction between different type of exits, that is, among merger and acquisition, split-up and various sorts of failure. Thus, a firm failure has remained an ambiguous focal point in innovation studies with one exception that is discussed next.

Innovation and failure

In contrast to innovation and firm success, some scholars have taken a different perspective and studied firm exits and failures. To our knowledge the only study linking innovations and business failure[2] was done by Barnett and Freeman (2001). Their results show that, while having innovative products prolongs the time of business activity, the simultaneous introduction of multiple products increases the mortality rates. Their argument bases on conception of organizational adjustment needed with innovation. When multiple products are commercialized at once, adjustments made for one innovation may complicate adjustments made for the others, especially if the products are closely related in their technologies and markets. Their results clearly illustrate that firm success does not solely base on developing a number of innovative products. In a worst-case scenario, innovativeness can lead to a firm failure if the pitfalls are not acknowledged. One of the pitfalls, according to Barnett and Freeman (2001), is the bad timing of commercializations.

Christensen *et al.* (1998) studied the rate of firm exit in a rapidly changing industry. Their results show that a transformative change in the competitive nature of the industry and managerial inability to adopt a radically developed state of the art leads to a higher probability of failure. Entering the industry a few years prior to emergence of a dominant design was also revealed as reducing the risk of failure. The results imply that it is crucial for firms to monitor industry changes and be able to identify major leaps in technology as well as adjust to them through innovation. Moreover, Thornhill and Amit (2003) found that external changes more often cause failure of older firms, while poor managerial capabilities led to downfall of younger firms rather than older ones. Thus, there seem to be environmental and managerial dimensions behind business success and failure, in addition to innovation characteristics and process-related factors.

Well-known studies on firm exits, on the other hand, have approached the issue on the industry level and have dealt with the matter of technological change as a broader phenomenon. Industry-level studies on firm survival and exits have focused on explaining the evolution of industries over time. The line of research has shed light on issues like the type of firms that enter and exit the markets, the effect that technology and the nature of industry have on this, and the relation of innovation and industry evolution. Tushman and Anderson (1986) examined major technological discontinuities and their influence on the competitive environment of firms. With their division of radical breakthroughs into two – competence-enhancing and competence-destroying – both were found to lower the entry-to-exit ratio in all studied industries. Especially in the case of competence-enhancing shifts in the nature of technology, resulting in greater product-class consolidation, more firms exit the markets. Particularly new firms face liabilities of newness and difficulties in keeping up with transformation based on previous knowledge possessed by incumbents. Thus, young firms often find it hard to adopt major innovations built on existing product know-how, and their probability to fail increases in the proceeding years of the breakthrough.

A comparable type of approach was taken ten years later by Klepper (1996), who stated that innovation and firm entry/exit relate to the technological life cycle of new products in an industry. Klepper concluded that in the beginning of a new, technologically progressive industry, the number of firms increases and the focus is on product innovations. Over time, process innovation is stressed and firms grow, eventually ceasing the entry of new firms. While the industry matures, producing firms compete on the basis of their size (scale economies) and their innovative ability. Furthermore, Agarwal and Audretsch (2001) argued that in different stages of industry's evolution, the size of the entrant as well as the nature of the industry make a difference in firm survival (see also Audretsch, 1995). They concluded that while large entering firms are more likely to survive in the formation stage of an industry, in the mature stage the size of the newcomer is less relevant. In addition to industry's maturity, capital intensity and subcontracting relationship have been found to be among industry-level factors promoting firm exit (e.g., Doi, 1999, in manufacturing industries).

Towards research on failing innovative firms

As previous research shows, the positive effects of innovation are various and they, as well as firm survival and failure, depend on: 1) characteristics of the innovation; 2) factors of the innovation process; 3) firm characteristics and management; and 4) business environment. The point of departure so far has been in, with differing perspectives, finding why some firms succeed, or survive. Innovation and renewal has been noted as a key issue for achieving business success, and thus revealing the ways to develop successful innovation has received a lot of attention. So far, the focus has been on portraying the right paths to success rather than getting more detailed information on the pitfalls. However, we argue that innovation does not invariably lead to success and growth. Some innovative companies fail to thrive, or even survive, despite the fact that the death ratio of innovative firms is lower than that of firms in general (see Saarinen, 2005). A number of innovators exit the markets also through mergers and acquisitions; whether they can be regarded as a success or failure is not relevant here. To this day, because of the fact that failure has been more or less the opposite of clear success, 'failure factors' have been typically conjectured as the ones *least* associated with *success* but not the ones *most* associated with *failure*. Furthermore, we find a lack of studies concentrating on firm failure factors on the level of innovation and innovation process.

We strive to extend the knowledge in the field of *innovation and failures* by bridging two thus far quite unlinked phenomena: innovation and firm failure. While the studies focusing on innovation projects determine failure as an innovation never to be commercialized (i.e., invention), we concentrate on firms with successful, commercialized innovations. The present study captures the success of an innovation project and links it with *ex post* failure at the firm level.

Defining 'failure'

In previous studies on innovation, firm survival and firm exit, failure firms have been seen as a part of exiting companies without any division between the two. Moreover, in some cases failures have been equated with market exits regardless the type of the exit. However, firm exit can even be the tiding of success, as Headd (2003) shows; one third of closed businesses are successful at closure. It can be a strategic manoeuvre to exit markets or sell your company to another one. Especially in the case of mergers and acquisitions, an exiting firm can be the most successful one rather than a failing one. A business can also be terminated because of personal reasons like retirement or ill health (Watson and Everett, 1996). Thus, it is justified to argue that an exit from the business register cannot be held as a business failure, but the reason for the exit has to be more thoroughly evaluated. As Saarinen (2005) illustrates, there have been more exits of innovative companies through mergers etc., than there have been natural exits (i.e., failures). In the present study, we follow a more veracious definition than in most studies and hold three reasons for exit indicating a firm failure; bankruptcy, liquidation and terminated business activity. We exclude mergers, acquisitions and split-ups from the reasons of failure. We use the Finnish business register, maintained by the National Board of Patents and Registration in Finland (PRH), to identify the failing companies among the recognized innovators.[3]

Data and analysis

Data

For our analysis, we draw on a unique dataset of individual Finnish product innovations. The database, named SFINNO,[4] provides micro-level insight into the wider processes of industrial renewal and technological change through individual innovations and innovation processes in Finnish businesses. The database currently includes the information of some 4500 individual product innovations, their innovation processes and of the approximately 1750 firms behind the development and commercialization of the innovations since 1945. Survey data on innovation processes is available in the database for innovations commercialized since 1985. The longitudinal database enables us to study innovations developed over a long period of time and use extensive information about the innovation process and the innovator firm in our analysis.

As outlined, our interest is on innovative, but failing businesses. We embarked on our data-gathering for this study by focusing on firms that have commercialized a product innovation during the years 1985–2005. For this period of time, we observe 2705 innovations in the SFINNO database and 1575 firms commercializing them. This group of companies are successful innovators, but not necessarily successful businesses. To distinguish between failing firms and non-failing firms, we apply the following approach.

To assess the age and life cycle of innovative and failing companies, we acquire the firm birth and death dates. The entry date of a firm is the date it was officially added to the business register. This date is used as an approximation for the time of business start-up. A similar method is applied in the case of exit date, which is regarded as the date of exit from the business register. The exit date is the point in time considered as equal with the time of a company closing down its business operations, that is, the approximated time of failure. The SFINNO database records also the commercialization year of innovations. Commercialization date is approximated as being 1 July for each innovation, resulting in a half-year tolerance margin either way, enabling us to calculate innovator's age with the accuracy of half a year.

Firm size is acquired from Suomen Asiakastieto Oy,[5] and it represents the size in number of employees at the time of commercialization of innovation, or where data was missing, we apply the preceding year of commercialization with employee data available.

To summarize our data collection, the SFINNO database is the basis for the data with innovation, innovation process and firm-level data. The business register was used to identify failures and to get data on their life span in more detail. Finally, firm data was completed with acquirement from the business information service. With the data, we aim at revealing the characteristics of firms that fail accordingly, in spite of introducing a novel product to the market. In addition, we examine characteristics of the innovations, and innovation processes. Thus, we distinguish between firm-, innovation process- and innovation-level characteristics.

Sample and descriptive analysis

We identified a group of 119 innovative firms failing before October 2007, which is 7.6 per cent of all companies that innovated in 1985–2005. Of the failing firms, 71 were found to be bankrupt, 37 failed through liquidation process and eleven exited markets and terminated their business for other reasons. In addition to our strict definition of firm failure, as bankruptcy is the most obvious indicator of business failure and the most frequent failure reason among the firms under study, our sample can be considered as representing the failure firms rather well. It is also interesting that fourteen of the failure firms were noted to have more than one innovation in the database,[6] equalling to 11.7 per cent of all failures. Six firms were observed to introduce two product innovations simultaneously, yet failing (supporting the findings of Barnett and Freeman, 2001). The survey data on innovations and innovation processes was available for about one third of the firms in our sample.

Because of time constraints, we selected a roughly equal number of non-failing firms from the SFINNO database. In order to account for the sectoral state of trade and the short-term economic climate, which can be reasonably assumed to affect the likelihood of business failures, the non-failing firms match the failing firms in terms of industry and year of commercialization. To verify the state of activity and to record the birth dates of these firms, we screened them with the help of the business register.[7]

In Figure 10.1, the failure rates of innovative firms are illustrated. The continuous line displays the failures under study by year of failure, while the broken line represents the total failure rate of all innovative companies in the SFINNO database.[8] When we look at the time of failures in the study, we notice that they have taken place most often in last decade. This is, we believe, because of two reasons. First, and most obviously, while our time frame is quite long (1985–2005) and the whole

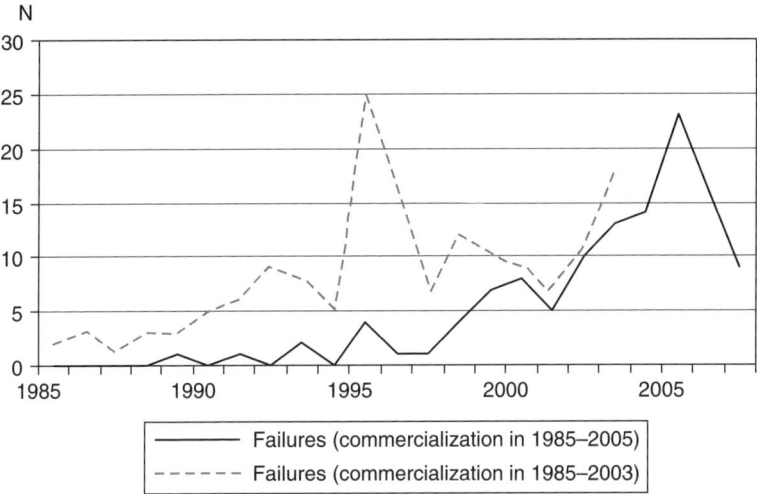

Figure 10.1 Failures of innovative firms (N = 119)[9]

period is well represented by having multiple innovations and innovators each year – firms by and large tend to survive a number of years. Second, the number of business failures in Finland has been generally increasing, and this is reflected in the rate of innovative firm failures as well (see also the broken line). The two peaks in Figure 10.1 are the consequences of economic crises; the first peak marks a severe recession at the start of the 1990s and the second of the IT bubble at the turn of millennium, both appearing with approximately a four- to five-year lag in failure rates.

Figure 10.2 describes the distribution of failure firms by age (in years) in three different phases of their life cycle. The dotted grey line displays the distribution by firm age at the time they started developing the innovation. The cohesive grey line displays the distribution by age at the time of commercialization of the innovation, and the black line by age at failure. We notice that 65 per cent of the failure firms[10] have started to develop their innovations at the same year they established the firm or before that (0 years). Firms have then commercialized their innovation at a rather young age. The new product was introduced to markets in the same year as start-up in 17 per cent of cases, and over half of the companies were at the age of three or under. Thus, many failing innovators are presumably entrepreneurial firms where the establishment of the

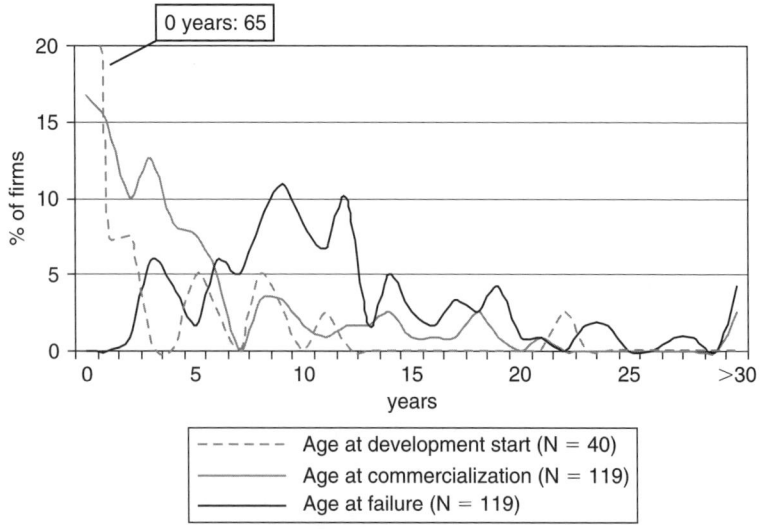

Figure 10.2   Life cycle of innovative failure firms

company and new product are tied together – established firms based on a new invention or technology. While the firms are in many cases young innovators, they do not seem to be failing as young, as frequently. Failing innovators are just under thirteen years old (12.8) on average at the time of business closure. A first peak in failures is at around three years of age, signalling firms that got a bad start and pulled their plug relatively soon after innovating. Most failures then occur between the ages of six and twelve, exactly 55 per cent, the median age of failures being ten years. The timeliness of innovation and firms' life cycle, illustrated by the three phases, raises two questions: is the innovation premium prolonging the survival of young firms especially (see Cefis and Marsili, 2006) and, moreover, is it the young, inexperienced innovators that fail more often at the end, and if so, why? This will be dealt with more in the results of the statistical analysis.

One advantage of the SFINNO database is that it also includes micro firms (i.e., firms with zero to nine employees). This is, for example, not the case in the Community Innovation Survey, which requires a minimum size of ten employees. It is important to include this category in failure studies, as small firms have been found to be more at risk (Evans, 1987; Dunne and Hughes 1994). When we look at the size distribution of innovative firms in our dataset, we find a different pattern for failing and non-failing innovators (Figure 10.3). The largest difference between failure and non-failure is found in the two extreme size groups. In the group of micro firms, failure appears to be much more prevalent, while less failure is seen in large firms.

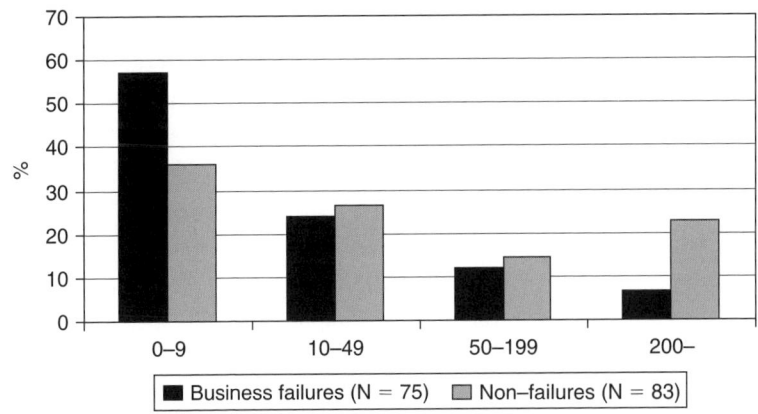

Figure 10.3 Size of innovating firms

Statistical analysis

Dependent variable

For our empirical analysis, our interest lies in how various characteristics on the firm, innovation process and innovation levels affect the presence of a business failure for innovative firms. We use a binary logistic regression model to estimate the effects of the x on the response probability:

$$P(y = 1|\mathbf{x})$$

Our binary dependent variable y here indicates, thus, whether or not any observed innovating firm has failed according to our previously described definition. We label this variable FAILURE, and set it to 1, where we observe a business failure. Our set of independent variables $\mathbf{x}$ denote the various characteristics $(\mathbf{x} = x_{i-j})$ of interest.

Explanatory variables

As the survey data was available for only a limited amount of cases, we chose to restrict the innovation- and innovation process-level variables and focused on some of the most appealing ones.

On the innovation level, we are primarily interested in aspects of the nature of the product innovation that firms introduced. The binary variable COMPLEX is constructed from a four-point complexity taxonomy developed in VTT Innovation Studies.[11] COMPLEX is set to 1 for highly complex innovations and 0 for innovations with a low complexity. High complexity here corresponds to Kleinknecht's definition (Kleinknecht and Bain., 1993, p. 44), where the innovation must be a system of many interrelated components stemming from different disciplines to fulfil our criteria. Our second innovation-level variable captures whether an innovation, or its underlying key technology, is protected by a patent or not. Thus, PAT is a binary variable set to 1 for cases in which a patent has been granted and 0 when not so.[12]

On the level of the innovation process, we have included the following variables in our model. To capture a possible role of governmental financial support, we incorporate the receipt of public funding for the development of the innovation by a binary variable which we label PUBSUB. This is set to 1 if any such subsidy was received, and 0 if not. Another, major characteristic of the innovation process is the pattern of collaboration. We also explore the potentially different effect of various types of partners for collaboration in the innovation process.

Therefore, we use CUSTCOLL to capture collaboration with customers, and SUBCOLL for subcontractors. These two variables are set to 1 if firms responded that they had collaboration with respectively customers or subcontractors in the innovation process, and 0 if not.

On the firm level, we use the following two variables to capture some key – characteristics.

INNOSIZE is a binary variable which indicates the economic dependence of the firm on the observed innovation. We constructed this variable from a five-point scale categorical variable indicating the share of the firm's total turnover that is generated by the observed innovation. We set our binary variable to 1 for cases of the highest class in the scale, namely when the innovation generates over 50 per cent of the company's turnover. The value 0 corresponds to shares anywhere between 0 and 50 per cent.

INNOAGE is a continuous variable which measures the age of the innovating firm in years (by one decimal) as measured from its start-up date to the time of commercialization of the innovation.

Results

Table 10.1 displays the results of our regression analysis. From our seven variables used in the analysis, only two appear to be useful in explaining the probability of failure of innovative firms. These two variables, INNOAGE and PUBSUB, are significant at the 0.10 level.

Both the firm age at the time of innovating and the receipt of public subsidy are negatively associated with business failure after innovating.

Table 10.1 Regression results

Variable	Coefficient	Significance
Innovation level		
COMPLEX	0,141	0,766
PAT	0,144	0,773
Innovation process level		
PUBSUB	−1,362	0,054
CUSTCOLL	−0,258	0,597
SUBCOLL	0,636	0,195
Firm level		
INNOSIZE	−0,100	0,848
INNOAGE	−0,790	0,050

Thus, more mature innovators seem less vulnerable to business failure. We also note that public subsidy has a greater effect in our model for the probability of failure than does the age of innovating firm. Firms that get public support for financing the development of innovation are more likely to elude business failure in the future. It is therefore interesting to consider here the mechanism of funding. In Finland, public funding for innovation is mostly received from Tekes (Finnish Funding Agency for Technology and Innovation). In evaluating the candidates for funding, 'High-quality, advanced technology and effective networking are essential criteria in Tekes' R&D funding decisions'.[13] This implies that public funding would enable firms to create high-quality, novel applications, which in turn would give them the competitive advantage which helps to prevent failure. Public funding also provides additional resources for innovators, which can even be substitutive for own funding. In this case, the risk of developing a new product is lower and the financial situation is more secured for a longer period of time. As a consequence, public subsidy prevents the firm from failing possibly even for a number of years after innovating and in various ways we do not observe here. Clearly, companies without public funding are more likely to fail.

The age of innovating firms (at the time of commercialization) has a negative effect on the probability of failure. This means that younger innovative firms appear to be more at risk to fail than their older counterparts. This outcome is not surprising and supports previous empirical findings in this matter (see Cefis and Marsili, 2005, 2006). It also seems to bear out our previous consideration of young innovators, namely that young, inexperienced companies can well fail in business, while succeeding in innovation. This finding raises an interesting question of what happens after innovation that leads to failure. Post-innovation performance seems to be critical for failure, and this is where young firms seem to run the most risk. This issue links to a study of Thornhill and Amit (2003), who found that young firms are more likely to fail due to deficiencies in general management and financial management. Our results indicate that this may also apply for innovating firms, especially in the era of post-innovation.

From our analysis, it seems that our studied characteristics of innovations do not have a demonstrable impact on the failure of firms. Neither the complexity, nor the patenting of innovations could be proven to matter for encouraging or eluding business failure. Somewhat surprising is that even though more complex innovations are regarded as harder and more resourceful to develop, thus putting more at stake, firms that have commercialized highly complex innovations do not seem to be more likely to fail.

A granted patent gives the patenting firm a temporary 'monopoly' for applying the technology in the markets and thus differentiates the product from potential rival products. Nevertheless, as limiting technological competition may lower the threshold of endeavouring in innovation, this protection mechanism does not seem to reach further by inhibiting business failures. Innovative companies that have protected their novel products are as likely to fail as non-patenting innovators. This implies that the positive effects of patent protection do not significantly overcome the negative factors leading to business failure.

The results for innovation-level factors are somewhat less surprising, as we previously observed that failing innovative firms do not fail in the subsequent years after innovation, but more often a number of years later. Thus, it is evident that the effects of single innovations for survival and failure decreases the longer the firm survives after commercialization. This implies that there is a certain innovation scope (see Georghiou *et al.*, 1986) that enables companies to capitalize on the innovation and its incremental improvements. When this scope ends, possibly with a new dominant design or other transformative change in the competitive situation, the failure risk increases. It can also be that after a successful first innovation the company is not able to learn and re-innovate, which is regarded as an important stage of the process of innovation management (see Tidd *et al.*, 2001). This is also implied by our data when the characteristics of the first innovation do not seem to have much influence on the scope for successive innovation.

Collaboration with the most obvious partners does not affect firm probability to fail. Both customer cooperation and subcontractor cooperation appear insignificant factors in explaining business failure of innovating firms. In particular the involvement of customers has been deemed to be a key success factor in innovation (e.g., von Hippel, 1988). In spite of being closely associated with the demand side of the innovation process and positively affecting marketization, effects from customer collaboration seem not sufficiently translated in sustainable business.

In addition, the relative size of the innovation for the firm does not seem to have an effect in our model. In other words, it does not make a difference whether the innovation is the primary product (>50 per cent of turnover) of the company or not. This is interesting, as one might expect a 'portfolio effect' to reduce the risk of doing business.

Conclusions

In this chapter, we studied the failure of innovative firms by bridging two thus far rarely linked phenomena: successful innovation and business failure. As the study is more exploratory in its nature, we applied a multilevel analysis in order to discover the factors behind failure and focused on the levels of innovation, innovation process and firm.

We found that failing innovative firms are more often young at the time of innovation, but do not typically fail in the subsequent years. The gap between innovation and failure is rather long and most failures occur around the age of ten years. In addition, and in the line of previous research, among innovative micro firms (under ten employees) failure occurs relatively more often, while the opposite is true for the large firms (over 200 employees).

With help of regression analysis, we observe that public subsidy has the greatest effect for preventing business failure. Firms that have received public funding for the development of an innovation, lowering the financial risks, are less likely to fail after commercialization. In addition, young innovative firms are more at risk to fail than older ones. Characteristics of innovation, collaboration with customers or subcontractors, or significance of the innovation for the firm do not seem to matter for the downfall of businesses.

The results suggest that, in order to avoid business failure, it is the post-innovation performance that matters the most, although the obtainment of public funding at the development stage of innovation is highly important. Firms seem to benefit from innovation for a number of years, but the scope of the first innovation will not last forever, and even multiple innovators can fail. Whether innovators are incapable of re-innovating, or poorly manage the era of post-innovation in general, remains an unanswered question.

Further research is needed in three issues. First, the effects of innovation and innovation process for the failure of firms need more detailed study, which is also our next goal. Second, the post-innovation performance requires a closer look in the case of failing businesses. In particular, the effect of repetitive innovation and continuous renewal is interesting, on the one hand, because of noted innovation scope and, on the other hand, because of the existence of failure even among multi-innovators. Third, micro firms form clearly an important research population for the present topic. Given their relative absence in empirical work, special attention on these firms is needed.

Limitations

Our data on innovative firms was restricted to consider the innovators of the years 1985–2005. While the used database has a wide coverage of innovations and the firms behind their development, we observed only 119 failing businesses. In addition, survey data on innovation processes and innovation characteristics was limited to cover around one third of our total dataset. Thus, our results are to be interpreted with these limitations in mind. It should also be borne in mind that this is the first study of its kind, and rather gives a potential direction for upcoming failure studies than final answers. However, it can be reasonably expected that, as we improve our data and model further by expanding our cases and improving measurement units in our data, the effects of age and public subsidy will be more pronounced.

Notes

1. For a more comprehensive literature review of success and failure of innovative projects, see van der Panne *et al.*, 2003.
2. Not only did Barnett and Freeman focus on failing innovative firms, they defined failure more strictly than just as 'exit', as well. They excluded firm's ownership – change events, where the business operations were not to continue, from failure definition.
3. We used the Finnish Business Register database which was updated at the end of September 2007. Criteria for a firm failure were that the firm had a 'ceased' status in the register and the reason for exiting the register of active firms was one of the three defined failure reasons. Because our data includes only limited companies, the used failure criteria (we also excluded cases with ownership changes in exit from our study group) can be held as signifying severe financial problems that led to firm failure. We also excluded a group of 51 inactive firms from failures with the help of PRH's experts. The inactives were companies that had been removed from the business register because they indicated no business activity for a longer period of time and did not respond to the requests of registry officials (e.g., did not send their financial statement in time or after requirements). Though the inactive firms had not indicated business activity recently, neither had they stated a termination of their business operations and thus could not be classified as failures according to our definition.
4. See http://www.vtt.fi/proj/sfinno/index.jsp for general information on the database, and Palmberg, 2004, Palmberg, 2006 and Lehto and Lehtoranta, 2006, for applications.
5. Suomen Asiakastieto Oy is a company providing business and credit information in Finland.
6. For these firms, we apply the data concerning the innovation that was commercialized first in our analysis. In most cases the earlier product introduction

was also regarded as more significant for the firm, as the second innovation was noted to be more or less a follow – up or a refinement of the first innovation.

7. As we realize that this procedure may introduce selection bias, we intend to expand the set of non – failing firms in the next version of our chapter, to cover the entire scope of the SFINNO database in the years of observation used in this study.

8. See also Saarinen (2005, p. 106) for historical development of exit rates of innovative firms.

9. Year 2007 includes failures from January until the end of September, while the data for the rest of the year was unavailable at the time of investigation.

10. Data on age at development start bases on SFINNO survey data and was available for 40 companies.

11. This taxonomy is a more applied version of the one used in Kleinknecht *et al.* (1993) (see, for details, Tanayama, 2002, and Saarinen, 2005).

12. As we are interested in the effect of protection, we do not consider patent applications as such in this case.

13. See http://www.tekes.fi/eng/tekes/rd/ for funding principles.

References

Agarwal, R. and Audretsch, D. B. (2001) 'Does Entry Size Matter? The Impact of the Life Cycle and Technology on Firm Survival', *Journal of Industrial Economics*, Vol. 49, No. 1, pp. 21–43.

Audretsch, D. B. (1995) 'The Propensity to Exit and Innovation', *Review of Industrial Organization*, Vol. 10, pp. 589–604.

Barnett, W. P. and Freeman, J. (2001) 'Too Much of a Good Thing? Product Proliferation and Organizational Failure', *Organization Science*, Vol. 12, No. 5, pp. 539–58.

Burgelman, R. A. and Sayles, L. R. (1986) *Inside Corporate Innovation: Strategy, Structure, and Managerial Skills*, New York: Free Press.

Cefis, E. and Marsili, O. (2005) 'A Matter of Life and Death: Innovation and Firm Survival', *Industrial and Corporate Change*, Vol. 14, No. 6, pp. 1167–92.

Cefis, E. and Marsili, O. (2006) 'Survivor: The Role of Innovation in Firms' Survival', *Research Policy*, Vol. 35, pp. 626–41.

Christensen, C. M., Suarez, F. F. and Utterback, J. M. (1998) 'Strategies for Survival in Fast-Changing Industries', *Management Science*, Vol. 44, No. 12, pp. S207–S220.

Cooper, R. G. (1979) 'The Dimensions of Industrial New Product Success and Failure', *Journal of Marketing*, Vol. 43, pp. 93–103.

Cooper, R. G. and Kleinschmidt, E. J. (1995) 'Benchmarking the Firm's Critical Success Factors in New Product Development', *Journal of Product Innovation Management*, Vol. 12, pp. 374–91.

De Brentani, U. (2001) 'Innovative Versus Incremental New Business Services: Different Keys for Achieving Success', *Journal of Product Innovation Management*, Vol. 18, pp. 169–87.

Doi, N. (1999) 'The Determinants of Firm Exit in Japanese Manufacturing Industries', *Small Business Economics*, Vol. 13, No. 4, pp. 331–7.

Dunne, P. and Hughes, A. (1994) 'Age, Size, Growth and Survival: UK Companies in the 1980s', *Journal of Industrial Economics*, Vol. 42, No. 2, pp. 115–40.

Evans, D. S. (1987) 'The Relationship Between Firm Growth, Size and Age: Estimates for 100 Manufacturing Industries', *Journal of Industrial Economics*, Vol. 35, No. 4, pp. 567–81.

Freeman, C., Robertson, A. B., Achilladelis, B. G. and Jervis, P. (1972) *Success and Failure in Industrial innovations: Report on Project SAPPHO by the Science Policy Research Unit*, London: Centre for the Study of Industrial Innovation, University of Sussex.

Georghiou, L., Metcalfe, J. S., Gibbons, M., Ray, T. and Evans, J. (1986) *Postinnovation Performance*, London: Macmillan.

Geroski, P., Machin, S. and van Reenen, J. (1993) 'The Profitability of Innovating Firms', *Rand Journal of Economics*, Vol. 24, pp. 198–211.

Goldenberg, J., Lehmann, D. R. and Mazursky, D. (2001) 'The Idea Itself and the Circumstances of Its Emergence as Predictors of New Product Success', *Management Science*, Vol. 47, No. 1, pp. 69–84.

Greve, H. R. and Taylor, A. (2000) 'Innovations as Catalysts for Organizational Change: Shifts in Organizational Cognition and Search', *Administrative Science Quarterly*, Vol. 45, No. 1, pp. 54–80.

Griffin, A. and Page, A. L. (1996) 'PDMA Success Measurement Project: Recommended Measures for Product Development Success and Failure', *Journal of Product Innovation Management*, Vol. 13, pp. 478–96.

Headd, B. (2003) 'Redefining Business Success: Distinguishing Between Closure and Failure', *Small Business Economics*, Vol. 21, No. 1, pp. 51–61.

Huergo, E. and Jaumandreu, J. (2004) 'How Does Probability of Innovation Change with Firm Age?', *Small Business Economics*, Vol. 22, No. 3. pp. 193–207.

Kleinknecht, A. and Bain, D. (1993) *New Concepts in Innovation Output Measurement*, New York: Saint Martin's Press.

Klepper, S. (1996) 'Entry, Exit, Growth and Innovation Over the Product Life Cycle', *American Economic Review*, Vol. 86, No. 3, pp. 562–83.

Lehto, E. and Lehtoranta, O. (2006) 'How Do Innovations Affect Mergers and Acquisitions? Evidence from Finland', *Journal of Industry, Competition and Trade*, Vol. 6, No. 1, pp. 5–25.

McDermott, C. M. and O'Connor, G. C. (2002) 'Managing Radical Innovation: An Overview of Emergent Strategy Issues', *Journal of Product Innovation Management*, Vol. 19, pp. 424–38.

Palmberg, C. (2004) 'The Sources of Innovations: Looking Beyond Technological Opportunities', *Economics of Innovation and New Technology*, Vol. 13, No. 2, pp. 183–97.

Palmberg, C. (2006) 'The Sources and Success of Innovations: Determinants of Commercialisation and Break-even Times', *Technovation*, Vol. 26, No. 11, pp. 1253–67.

Rothwell, R., Freeman, C., Horsley, A., Jervis, V. T. P., Robertson, A. B. and Townsend, J. (1974) 'SAPPHO Updated-Project SAPPHO Phase II', *Research Policy*, Vol. 3, pp. 258–91.

Saarinen, J. (2005) 'Innovations and Industrial Performance in Finland 1945–98', *Lund Studies in Economic History*, Vol. 34, Stockholm: Almqvist & Wiksell International.

Saarinen, J. and Niininen, P. (2000) 'Innovation and the Success of Firms', VTT Working Papers, No. 53/00, VTT, Espoo.

Saarinen, J., der Have, R. P. and Pesonen, P. (2007) 'Sfinno: Database of Finnish Innovations: Data Overview', *2007 Kauffman Symposium on Entrepreneurship and Innovation Data*, SSRN database, accessed 15 December. Available at: http://ssrn.com/abstract=1022688

Schumpeter, J. A. (1934) *The Theory of Economic Development: An Inquiry Into Profits, Capital, Credit, Interest, and the Business Cycle* (third printing, 1963), New York: Oxford University Press.

Tanayama, T. (2002) *Empirical Analysis of Processes Underlying Various Technological Innovations*, VTT Publications, No. 463.

Teece, D. J., Pisano, G. and Shuen, A. (1997) 'Dynamic Capabilities and Strategic Management', *Strategic Management Journal*, Vol. 18, No. 7, pp. 509–33.

Thornhill, S. and Amit, R. (2003) 'Learning About Failure: Bankruptcy, Firm Age, and the Resource-Based View', *Organization Science*, Vol. 14, No. 5, pp. 497–509.

Tidd, J., Bessant, J. and Pavitt, K. (2001) *Managing Innovation: Integrating Technological, Market and Organizational Change* (2nd edn), Chichester: John Wiley.

Tushman, M. L. and Anderson, P. (1986) 'Technological Discontinuities and Organizational Environments', *Administrative Science Quarterly*, Vol. 31, No. 3, pp. 439–65.

van der Panne, Gerben, van Beers, Cees and Kleinknecht, Alfred (2003) 'Success and Failure of Innovation: A Literature Review', *International Journal of Innovation Management*, Vol. 7, No. 3, pp. 309–38.

Watson, J. and Everett, J. E. (1996) 'Do Small Businesses Have High Failure Rates? Evidence from Australian Retailers', *Journal of Small Business Management*, Vol. 34, No. 4, pp. 45–62.

11
Framing Elements of Service Innovation

Robert van der Have

Introduction

Apart from having a dominating share of GDP in the developed, present-day economic landscape (Preissl, 1997), services have also been highly associated with economic growth (Kravis *et al.*, 1983). Apart from their share, services thus form an increasingly important part of economic activity for further economic advancement. Moreover, there is a growing consciousness about the critical role that services play in the advancing development of the manufacturing sectors. First, many manufacturing firms regard the provision of services as a key growth area for their businesses nowadays. Second, the importance of knowledge-intensive service activities has been found to be significant in industrial development (Hipp and Grupp, 2005; Leiponen, 2001; OECD, 2006; Tomlinson, 1997).

Given the considerable and broad impact of these trends, it is therefore surprising that the amount of attention for innovation in services is still rather much unequally represented in studies of innovation by firms. Historically, the amount of attention on technological innovation has been dominating. However, a reverse trend can be noted in recent times, witnessing a broader growing discussion among academics and practitioners around the debated need for a 'Services Science'.[1] Yet, much work is needed to improve our relative lack of understanding of the sources and characteristics of innovation in services. As Tether (2005) argues, conceptual models together with empirical approaches are in need of being broadened, as their capturing of service innovations is still inadequate. Eventually, better-informed empirical studies will be able to analyze a broader range of firms and products, improve our current understanding of the characteristics of service innovations and their underlying innovation processes.

The aim of this chapter is therefore to contribute to this ongoing development by reconsidering the conceptual framework developed by den Hertog and Bilderbeek (1998, 1999), and systematize a number of findings in extant services literature, as well as two alternative service models. Developing den Hertog and Bilderbeek's conceptual framework on the loci of innovation further on the basis of these insights can help to improve our ability to recognize innovations in services in empirical work. The present chapter can be viewed as part of a logical next phase in innovation studies with a micro-level approach at large, namely by taking an integrative, or *synthesizing* conceptual approach (see also Drejer, 2004). That is, the contrasting specificities to services as well as their similarities as compared to innovation in manufacturing are drawn together. This means here that both tangible and technological aspects as well as the often non-tangible, 'softer' aspects of innovations – which have been found to be of importance in service sector-specific studies – have a place in the framework. Thus, the increasing intertwining of services and manufacturing activities is also recognized. With clearly challenging tasks ahead for students of innovation, a need exists for further conceptual development, which takes the peculiarities, complexity and diversity of service activities into account in this integrative manner.

Therefore, this chapter is intended to support the *study* of service innovations and their innovation processes. It is useful to depart from a 'system notion' of services, which includes both the tangible and intangible parts of services, being intrinsically *synthesizing*. Thus, it is of crucial importance in this study to utilize a conceptual service innovation model not refrained from the relevant focal areas of SEM (Service Engineering and Management), as its more extensive notion of services has much to provide to service innovation studies. To this end, the above-mentioned 'synthesizing' conceptual approach to service innovation connects the here-developed research framework for innovation in services and the broader SEM framework. More especially, as quality improvement is also an important objective in SEM, added value as perceived by customers not only stems from artefacts (which, for example, serve to deliver and automate services), but involves frequently intangible means of value creation, for example, in the sphere of knowledge and service contents (Goldstein *et al.*, 2002).

The remaining parts of this chapter will, first, provide a definition of a service product and on the basis of this a definition of a service innovation. Next, I will briefly discuss two of the few alternative conceptual models that are available in the services literature and meaningful to study innovation in services. These are the models developed by Gallouj and Weinstein (1997) (the 'Lille model'), and Edvardsson and Olsson (1996) (the 'Nordic model'). After this, the conceptual model of service innovation by den Hertog and Bilderbeek will be introduced. This model will then be described in further detail with the aim of clarifying what can be changed in a service offering by means of innovation. Along with describing the model, some additional insights from the service innovation literature as well as the former two models will be integrated. Finally, the chapter ends with a brief concluding discussion.

A definitional underpinning

Before exploring which aspects and characteristics of service innovations should be incorporated, it is useful to first of all provide an overall definition of what we first of all understand to be 'service products' (as product offerings) and, following, 'service innovations'.

Based on interpretations of (Gadrey *et al.*, 1995) and recent literature reviews conducted by (de Jong *et al.*, 2003; Edvardsson *et al.*, 2005), the following working definition of a service product is here proposed as a point of departure:

> A delivery of a solution to solve a problem or fulfil a need in the form of an activity (or series thereof), which goes beyond a mere supply of a tangible good for that purpose.

Consequentially, at least a part of the 'solution' is of an intangible nature and stems either directly or indirectly from human action (be it intellectual, speech, writing or physical performance). This can be in the form of help, utility or care, as well as information, experience, expertise or knowledge.

The mentioned intangible component of a service product and the centrality of activity in its definition, then, have strong implications for defining a 'service innovation' as well. Namely, a service innovation may constitute a non-technological change, such as a change in the stages of a process or intensifying a provider–client relationship. Based on the above and de Jong *et al.* (2003), we perceive service innovation as:

> The introduction of, additions to, or changes in service products, their concept, and/or the process that generates the service product which are new to the service providing organization and its clientele or/and the market it operates in.

This definition encompasses a) a centrality of change and renewal, b) allows for intangible dimensions of innovation and c) excludes any form of change or renewal which has been rejected or not adopted by the producer or clientele. In this latter case, the change has not passed the stage of 'invention'. Therefore, it should not be counted as an 'innovation' as it does not lead to either market adoption or other forms of internal utility for the producer in practice, and thus the creation of value.

Evaluating prospective models of service innovation

A next step, then, is to describe which non-tangible and possibly tangible elements of service innovation can be distinguished. In the introduction to this chapter, it was pointed out that, besides artefacts as objects of innovation, non-tangible elements of innovation need to be considered in service innovation research as well. Otherwise, a risk exists that those elements in innovations will be overlooked and misunderstood. Doing so then opens up interesting and highly relevant research avenues as to which factors in the innovation process might play a salient role to innovation in these various dimensions. However, caveats exist in that identifying a service innovation may in some cases be more ambiguous and challenging as a result. This is caused by the less clear-cut boundaries to and within service products, for which product, process and consumption are often perceived to fall together at least to some extent (OECD, 2005; Gallouj and Weinstein, 1997; Edvardsson *et al.*, 2005). It is therefore all the more crucial to find a model which is not too ambiguous in pin-pointing the loci of innovation in service products.

The model focusing on characteristics

This model, which was developed by (Gallouj and Weinstein, 1997), belongs to the so-called 'Lille school' of service studies. Like the model by den Hertog and Bilderbeek, this conceptual framework models a service with the intention to study innovation in services. However, what distinguishes the Lille model from that of den Hertog and Bilderbeek is that it is based on (but further developed from) Lancaster's (1966) work on product characteristics, which was originally applied to manufactured goods. It consists, therefore, of a set of three distinct technical (internal) and service (external) characteristics (Gallouj and Weinstein, 1997): a) the final characteristics, which are seen from the end-user perspective. Final characteristics describe which kind of service the product renders to its user, such as its utility and performance; b) the technical characteristics of the product, which are perceived from the product itself (hence an 'internal characteristic'). This element describes therefore the means by which the final characteristics are created, such as the tangible (e.g., 'hard' technical factors) and intangible (e.g., organizational aspects, or codified knowledge) systems that produce the service, as well as processes; c) a service's competence characteristics refer to the sets of involved human skills from the service provider/producer and the customer, which are mobilized by the technical characteristics in order to produce and deliver the service (Gallouj and Weinstein, 1997). In this model of a service product, an innovation is defined as a change affecting any of the three characteristics.

An advantage of this representation, as mentioned by the authors themselves, is that the problematic manufacturing-based distinction between product and process innovations is dropped. This is because the model treats processes as part of the service product. One other advantage is that the model integrates intangible elements of a service product, such as organizational aspects. The model has, however, some drawbacks as well. As Toivonen *et al.* (2006) note, the category of 'technical characteristics' is rather broad and flexible. Consequentially, the model is not well able to pin-point more concrete parts of service products, which is a prerequisite to identify service innovations in empirical studies. Also, the 'final characteristics' are more the consequential outcome of the other two categories, and thus not a locus of innovation in itself. Therefore, this model seems not very useful for our purposes here.[2]

The model of prerequisites

The service model developed by Edvardsson and Olsson (1996) and Edvardsson (1997), like the former model, starts with the customer's qualitative perception of the outcome of a service product. The model is much less aimed at pin-pointing loci of service innovations, but serves to explain which prerequisites a service must have in order to achieve a sufficient outcome for the customer. Therefore, the so-called 'Nordic model' is not explicitly linked to product innovation by the authors.

This model, with its background in marketing, is aimed to support the development of a new service by a service provider, and is therefore still worth considering here, given its relative proximity to both innovation in services and SEM.

The first-distinguished and crucial part in this model is the *service concept*. This is defined as a detailed description of *what* the needs of the customer are, along with a description of *how* they are met. The concept includes, therefore, both content and structure of the service, but on a somewhat general level.

Second, the authors distinguish the *service system*. This dimension describes the whole set of resources needed to produce the service. These resources are grouped as distinct subsystems: a) the service company's workers (human resources that embody intangible aspects such as knowledge and skills); b) the customers (for example, customers can provide information, perform administrative tasks or use some equipment in order to co-produce the service); c) the physical/technical environment, which include premises, technical systems or equipments at partners' or customers' domains; and d) the last subsystem, 'organization and control', includes the organizational structure, administrative support systems (e.g., financial or information systems), but also the issue of how interaction with the customers is arranged (e.g., how feedback is arranged, as well as opening hours and loyalty programmes) and, finally, marketing tasks.

The final part of a service is the *service process*, which refers to the chain of (possibly parallel) activities that produce the service. Activities can also take place at partners' or customers' domains.

In spite of not being intended as a model for the loci of service innovation, the model is in some areas more concrete than that of Gallouj and Weinstein. A number of parallels can also be observed: the service concept shows analogy with Gallouj and Weinstein's 'final characteristics' in that it describes the benefits (the 'what') for the customer. However, the strength of Edvardsson's model is its addition of the 'how' question in the concept, which is important to include when we study a new service product. Gallouj and Weinstein's 'technical' and 'competence' characteristics can also be recognized in this model's service system, which makes these resources actually more concrete.

On the other hand, the model also shows some problems in studying innovation in service products. Namely, it goes quite a bit further than the service product as such, and arguably looses touch with the service product to a problematic extent. This is visible in the element 'organization and control', where some firm-level operational systems and tasks are included which do not necessarily form part of the service product itself. Notwithstanding the potential importance of the organizational element in service products – such as the organization of customer interaction – it is better for our purposes here to keep a direct link to the service product. In the context of service product innovation, the inclusion of customers as a resource or element is somewhat problematic as well. In a strict interpretation this means that changing customers (e.g., a new market segment) would lead to product innovations (i.e., the product is altered) rather than market innovations. This is not necessarily always the case, for example, in highly standardized services. Therefore, it seems better here to not go beyond the organization and interface of interaction with the customer.

A strong point is that the model allows for intangible aspects to services, such as organization and skills. However, here the model is too broad for our purpose: not all, but only those embodied skills and knowledge directly involved in producing the services are relevant. Finally, the model has one important advantage besides including intangibles: it recognizes that a service product can exist beyond organizational boundaries. This is made clear by the fact that technical systems or equipments at partners' or customers' domains can be part of producing one integrated service product.

The above discussion of two service models highlights that a good model for service innovation should be concrete enough about the elements of a service product which can be innovated. The model should also maintain a good balance of organizational breadth (i.e., to look beyond the organizational boundaries of the service-providing firm), on the one hand, and, on the other hand, at the same time keep its focus fairly tightly centred on the service product.

The model of the loci of innovation

The final model under discussion here, and chosen for further development, deals explicitly with loci of innovation in services, and was originally developed by den Hertog and Bilderbeek (1998, 1999). It shares the advantage of the previous models in dismissing the exclusiveness of the tie between innovation and technology and addressing unambiguously non-technical and intangible dimensions of innovation. The reason why it is here selected to develop further is two-fold. In line with the previous models, is its multi-dimensionality (i.e., containing as set of multiple loci), which offers the possibility of mapping the systemic nature of an existing service product. This allows the study of possible interdependencies among various dimensions which can become salient when an innovation occurs in one dimension. Second, and more distinctively, the model is more tightly centred on the *service product* but nevertheless allows us to consider innovation at different levels, such as the (inter-)organizational level.

The model – in its extended/elaborated version – is depicted in Figure 11.1. By drawing on the contributions made by the two previously discussed models, the model below can be enhanced by improving the concreteness in its loci of innovation (depicted in the boxes in

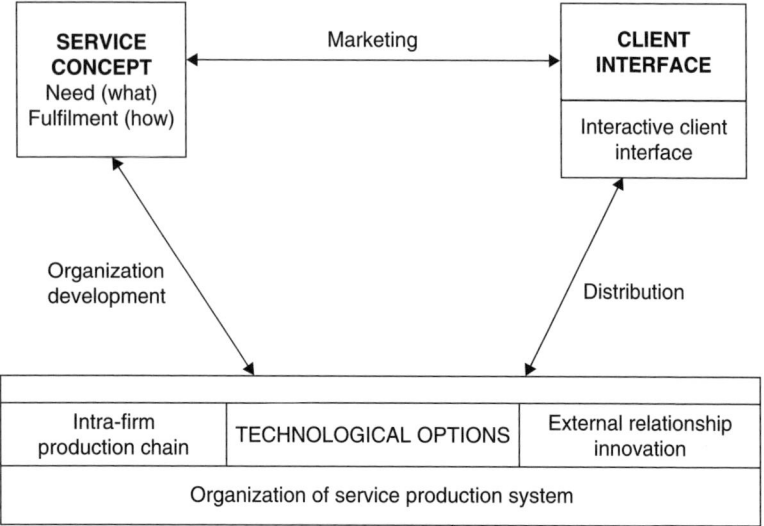

Figure 11.1   Conceptual model of the loci of service innovation
Source: Adapted from den Hertog and Bilderbeek, 1998.

Figure 11.1). For the various loci, the text in capitals is unaltered from the original model. Next, I will address each locus in turn.

The *concept* element is originally not made very explicit. However, den Hertog and Bilderbeek (1999, p. 6) mention illustratively that it refers to 'a new idea or concept how to organize a solution to a problem'. But this can be made more concrete by following the Nordic model in which the concept of a service constitutes the basic idea of *what* needs or wishes are addressed for a customer and *how* this is carried out. In other words, it constitutes the rudimentary service that is provided to a client, and its elementary content and structure as seen from the provider (Goldstein *et al.*, 2002). Change in this element of services has, therefore, far-reaching consequences for other elements of the service on a more detailed level as well. For analytical clarification, 'concept' is here limited to the service product offer itself in order to denote a service innovation of this kind. The concept element comes once again very close to what Gallouj and Weinstein (1997, p. 539) call 'the final (or use) characteristics of the good', as seen from the end-user's point of view. However, the term 'final characteristics' is perhaps slightly narrower as it does not necessarily include any broad definition of what the provider does.[3] A key point is that a concept must be new in its application within a particular market in order to denote an innovation in a service product (den Hertog and Bilderbeek, 1999). The *client interface* dimension relates to the way a service activity connects to the client (i.e., the means of linkage). This often comprises a medium or communication channel, but can also have the form of changes in physical presence, like the introduction of the ATM in the banking sector (den Hertog and Bilderbeek, 1998, 1999). A proposal made here is to elaborate this dimension with changes to the way of organizing the process of interrelating with the client (e.g., see Howells and Tether, 2004), as this is also a way of linkage which is related to the service product, but with an *interactive* nature. The literature repeatedly indicates the importance of clients as co-producers of both services and service innovations (e.g., Hipp and Grupp, 2005; Edvardsson and Olsson, 1996; Toivonen and Tuominen, 2009). Therefore, it is useful to distinguish a sub-component, which is here labelled *interactive client interface* in order to differentiate it from static, one-way interfaces. As such, the latter element comes close to what Edvardsson and Olsson (1996) name 'customer process', which is part of the service-generating process where the customer takes part as co-producer. Therefore, changes in the role, participation and responsibility of the customer in the client-provider relationship (Edvardsson and Olsson, 1996) can be qualified as innovations in the interactive client interface.

The *organization of service production system* here has a stricter representation than that given by the original authors (den Hertog and Bilderbeek, 1998) of the basic model defined (i.e., 'New Service Delivery System'). Here, as an element of service innovation, the delivery part of the production belongs analytically to the previous category.[4] An important part of this dimension is that it involves the organizational design structure for producing services. Denoting innovation in this sub-part can involve issues such as reconfiguring, adding or eliminating parts of the service production system. Gallouj and Weinstein (1997) have made some very strong proposals to analyze the nature of service innovations.[5] Partly following Djellal and Gallouj (2001) and Drejer (2004), a salient subset will be distinguished here, labelled *external relationship innovation*, 'defined as the establishment by a firm of particular relationships with partners (suppliers, public authorities or competitors)' (Drejer, 2004, p. 558). However, the focus is limited here on the organization of service production process or chain, thus maintaining a direct link to the service product as an innovation.[6] Also, as the customer's side is addressed separately above due to the special nature of the demand-side involvement in the production process, external relationship innovation excludes the client side here conceptually.

The final distinguishing element of a service product in which innovation may take place here refers to pure technical change: the *technological options*. In the original version, this locus was separate from the service delivery system, and placed central in the model. A new technology can drive a new service, an existing service provision can assimilate a new technology for doing so, or service providers can apply existing technologies in novel ways, for example, by customization or finding new purposes for it. A service provider can also acquire new technology for internal use of producing a service, for example, by introducing a new information system as a resource. Thus, this dimension includes changes to technologies which are used to both produce and deliver services. It is worth noting here that this possibly coincides with the second locus (client interface), and forms part of the service production system. In the original model, it seems artificial to separate it from the delivery (here: 'production') system, but it is still useful to distinguish as a locus: *denoting* technical changes separately in a service innovation model has the benefit of enabling observation of changes where only a technical resource is changed, or in which this impacts also other resources (e.g., human resources in the form of requiring new skills by means of input) or the organizational and process aspects in the service. This is also better in line with the SEM perspective on services as systems and therefore relevant to our objective if we want to study service innovation not refrained from this.

This extended service innovation model can help to denote or disentangle the loci of innovation in services. However, due to the nature of service activities – in which production and consumption may co-occur simultaneously in one and the same process – an innovation may constitute more than one dimension simultaneously. I regard this not as a problem, but as an opportunity to address the complexity of service innovation in future empirical work by studying interrelationships of various kinds. The relevance of this is underlined by the broader and integrative SEM approach to services.

Conclusion

It is clear that research into service innovations is far from mature and has room for advancement both on the conceptual and on the methodological side. This chapter has focused on the former. Service innovation studies is a field with many challenges left. By addressing and connecting complementary approaches to services, and its conception of a service as being a system with diverse aspects (e.g., organizational, technical and interactive), the SEM approach seems a potentially useful source of inspiration for service innovation students. This chapter contributes by taking an incremental step in this direction.

The aim of this work has been to systematize some relevant loci of service innovation in service products and synthesizing them into a reassessed conceptual model earlier developed by den Hertog and Bilderbeek (1998). This is necessary in order to deal with the heterogeneity of what service products can comprise, and the variety in which they can be renewed. Moreover, the framework shows that services are in fact in many cases sets of interdependent, interrelated and quite widely ranging elements. Therefore, in service innovation studies – as in SEM – service products can be treated as systems products as well. It is not feasible to completely separate or isolate its individual elements, but we are at least able to denote changes in any of these, even as change in its elements may in reality co-occur.

One important feature of the conceptual model by den Hertog and Bilderbeek (1998) has come forward through the critical discussion above. Namely, we see that, due to the system and process aspects of services, changes in service products can also involve changes in the organizational and networking (suppliers, customers, etc.) domains. The ability to indicate innovations in these different domains is a particular strength of this model. However, the focus has here remained on the product level (i.e., the different domains had to be considered as being a real part of the service product). Yet, this issue also exposes the need to consider innovation taking place on the level of organizations and markets, as there are clear connections, which ask for more analytical work. Future conceptual work could assess the possibility to extend the model in this direction, for which it seems to have potential grounds. Nonetheless, this is beyond the scope of this chapter.

Finally, the existing array of empirical data on service innovation is – apart from case studies – mostly acquired by surveys only. This provides limitations in dealing with the complexity of services among others (cf. Tether, 2005; Sirilli and Evangelista, 1998). By aiming to collect data from the specific innovation (the so-called 'object-based approach') as point of departure (and articulating concrete loci or elements of service innovation), not only does more information become available for the researcher, but also problems of subjectivity for interviewees and questionnaire respondents are more restricted. A model which is capable of capturing changes in the organizational and networking sphere, therefore, is also particularly useful to reduce ambiguity when one wants to collect more additional data to the object-based method in the form of additional surveys or interviews directed at the innovators.

Notes

1. See the recent special issue of Communications of the ACM, 49(7), July 2006.
2. Though the authors provide strong proposals for the studying of the nature of innovation in services (see, for a broader assessment, van der Have *et al.*, 2007)
3. The 'final characteristics' element of a service product is in their notion more biased to its customer utility.
4. Note that, in reality, a continuum may exist within a service offering, in which production and delivery are intertwined. However, it should be kept in mind that conceptual separation supports recognition of an innovation, and the present model is not intended as a service model, but rather a service *innovation* model.
5. This discussion is taken up in van der Have *et al.*, 2007.
6. Bearing in mind the frequent intertwining of product and process in services (Sirilli and Evangelista, 1998).

References

De Jong, J. P. J., Bruins, A., Dolfsma, W. and Meijaard, J. (2003) 'Innovation in Service Firms Explored: What, How and Why? Literature review', strategic study report B200205 in research programme SMEs and Entrepreneurship, EIM Business and Policy Research.

den Hertog, P. and Bilderbeek, R. (1998) 'Conceptualizing (service) Innovation and the Knowledge Flow between KIBS and their Clients', SI4S topical papers, No. 11, Oslo: STEP Group, Studies in Technology, Innovation and Economic Policy.

den Hertog, P. and Bilderbeek, R. (1999) 'Conceptualising Service Innovation and Service Innovation Patterns', thematic essay within the framework of the Research Programme Strategic Information Provision on Innovation and Services (SIID) for the Ministry of Economic Affairs, Directorate for General Technology Policy.

Djellal, F. and Gallouj, F. (2001) 'Patterns of Innovation Organisation in Service Firms: Postal Survey Results and Theoretical Models', *Science and Public Policy*, Vol. 28, No. 1, pp. 57–67.

Drejer, I. (2004) 'Identifying Innovation in Surveys of Services: A Schumpeterian Perspective', *Research Policy*, Vol. 33, No. 3, pp. 551–62.

Edvardsson, B. (1997) 'Quality in New Service Development: Key Concepts and a Frame of Reference', *International Journal of Production Economics*, Vol. 52, No. 1, pp. 31–46.

Edvardsson, B. and Olsson, J. (1996) 'Key Concepts for New Service Development', *The Service Industries Journal*, Vol. 16, No. 2, pp. 140–64.

Edvardsson, B., Gustafsson, A. and Roos, I. (2005) 'Service Portraits in Service Research: A Critical Review', *International Journal of Service Industry Management*, Vol. 16, No. 1, pp. 107–21.

Gadrey, J., Gallouj, F. and Weinstein, O. (1995) 'New Modes of Innovation: How Services Benefit Industry', *International Journal of Service Industry Management*, Vol. 6, No. 3, pp. 4–16.

Gallouj, F. and Weinstein, O. (1997) 'Innovation in Services', *Research Policy*, Vol. 26, No. 4, pp. 537–56.

Goldstein, S. M., Johnston, R., Duffy, J. and Rao, J. (2002) 'The Service Concept: The Missing Link in Service Design Research?', *Journal of Operations Management*, Vol. 20, No. 2, pp. 121–34.

Hipp, C. and Grupp, H. (2005) 'Innovation in the Service Sector: The Demand for Service-specific Innovation Measurement Concepts and Typologies', *Research Policy*, Vol. 34, pp. 517–35.

Howells, J. and Tether, B. (2004) *Innovation in Services: Issues at Stake and Trends*, Brussels: Commission of the European Communities.

Kravis, I. B., Heston, A. W. and Summers, R. (1983) 'The Share of Services in Economic Growth', in F. G. Adams and B. G. Hickman (eds), *Global Econometrics: Essays in Honor of Lawrence R. Klein*, Cambridge, MA: MIT Press.

Lancaster, K. J. (1966) 'A New Approach to Consumer Theory', *Journal of Political Economy*, Vol. 74, No. 2, pp. 132–57.

Leiponen, A. (2001) *Knowledge Services in the Innovation System*, Helsinki: Taloustieto.

OECD (2005) *Promoting Innovation in Services*, Project Report DSTI/STP/TIP(2004)4/FINAL on the OECD project 'Enhancing the Performance of the Service Sector', Paris: OECD.

OECD (2006) *Innovation and Knowledge-Intensive Service Activities*, Paris: OECD.

Preissl, B. (1997) 'Services in Europe: Patterns of Growth and Development–Statistical Overview', SI4S-WP2/1997 Survey Report.

Sirilli, G. and Evangelista, R. (1998) 'Technological Innovation in Services and Manufacturing: Results from Italian Surveys', *Research Policy*, Vol. 27, No. 9, pp. 881–99.

Tether, B. S. (2005) 'Do Services Innovate (Differently)? Insights from the European Innobarometer Survey', *Industry and Innovation*, Vol. 12, No. 2, pp. 153–84.

Toivonen, M. and Tuominen, T. (2009) 'Emergence of Innovations in Services', *The Service Industries Journal* [online], Vol. 29, No. 5, pp. (2009).

Toivonen, M., Tuominen, T. and Brax, S. (2006) 'Innovation Process Interlinked with the Process of Service Delivery: A Management Challenge in KIBS', paper presented at the conference 'Innovation in Services', 15–17 June, Manchester, ESRC Centre for Research on Innovation and Competition (CRIC), University of Manchester.

Tomlinson, M. (1997), 'The Contribution of Services to Manufacturing Industry: Beyond the Deindustrialisation Debate', CRIC Working Paper No. 5, Centre for Research on Innovation and Competition (CRIC), University of Manchester.

van der Have, R. P., Toivonen, M. and Tuominen, T. (2007) 'Dimensions of Service Innovation', paper presented at the 5th International Symposium on Management of Technology, ISMOT '07, Vol. 1–3 June, Hangzhou, China.

Part V
Epilogue

12
Studies of Innovation: Challenges and Possibilities – The Researchers' Perspective

Nina Rilla and Pekka Pesonen

What is innovation all about?

The importance of innovations for long-term growth on economic and firm level has been widely acknowledged (Cantwell, 1999). However, the explanation of the term 'innovation', the main motor of industry and company renewal, has often been neglected. Moreover, the lack of common definition has been a hindrance, for instance, in making comparative studies (Johannessen *et al.*, 2001). In order to explain innovation, we need, first, to make a distinction between invention and imitation. Invention is an idea or a model for something (Freeman, 1987), whereas in order for invention to be called an innovation, it has to be commercialized on the market by a business or equivalent (Schumpeter, 1963; OECD, 2005), that is, to be diffused to other parties beyond the discoverers (Garcia and Calantone, 2002). Imitation, on the other hand, is an innovation which has been copied by others (Marquis, 1988). Imitation has not by any means always been regarded with a negative connotation but has been highly appreciated activity; this is still the case today, since imitation can also be understood as technology diffusion.

Innovation incorporates a couple of central elements: first, innovation needs to provide newness to someone, whether to developer or user (Rogers, 1982). This makes diffusion of innovation a central dimension of economic progress as well (Godin, 2008). The second important aspect is activity, in particular learning – an ability to take advantage of previously developed elements. And, third, it provides commercial benefit for someone (Tidd *et al.*, 1997). When we think about innovations today, and how we understand them, a large share of innovations are built on already-existing inventions, or rather a combination

229

of inventions (Schmookler, 1966), or on reconfiguration of prevailing products (Henderson and Clark, 1990). Innovations are, most of all, created through communicating and combining of knowledge (Kogut and Zander, 1992). Innovation, as the word implies, is something that is needed for renewal. In the firm context, we usually refer to a new and improved commodity, which is realized, for example, in the form of product, service or process. In Schumpeter's words, innovations are 'new combinations', which drive the economic development as they challenge the old existing products, technologies and companies (Schumpeter, 1963; Drejer, 2004). This is to say that innovations should not simply to be understood as tangible objects, but more crucial is their novelty characteristic. Novelty brings renewal, which enables companies to stay competitive and economy to flourish.

Innovation process

In order to get a grasp of the technological change (Nelson and Winter, 1982) and progress, that is, the use of technological inventions in industrial processes, the innovation process thinking arose in the 1950s. The management- and business-oriented scientists among economists started to be interested in technological change and its sequential development, in which innovation was pictured and analyzed as a process. Innovation was realized through organizations and mobilizing employees' creative abilities (Godin, 2008). Therefore, it was important to acknowledge that innovation is not a static phenomenon but a combination of various actors and activities. Solving of the mystery of 'the black box', as the innovation process was largely seen, became an interest of various disciplines (Fagerberg, 2006).

From the 1950s till the 1970s, the innovation process was regarded as linear process, which evolved from idea to commercialized innovation. At first, the emphasis on science and technology as the initiators of innovation dominated the thinking, and pictured innovation process as a technology-push model. After the war period, the market and demand perspective raised its head due to high concentration on industrial and economic growth, which led to development of the market demand model. Increased interest in innovation studies in the 1970 and 1980s gave room for more advanced and complex innovation process models. These new, more holistic models stressed the integration of various activities, and incorporated another important aspect in innovation, the feedback loops between activities. Integration is still a dominant characteristic in today's innovation process studies (Rothwell, 1994).

Studying changes in innovation process

Innovations are developed in complex processes and they depend on a variety of factors. In order for innovation to succeed it requires sufficient conditions not only inside the firm responsible of the development but also in the external environment, which should be supportive and beneficial for innovation. The technological knowledge and know-how, which provides roots for innovation to evolve, are often a result of long-term and large investments made in developing know-how and conditions for certain technologies to flourish. The development depends most of all on companies' capabilities to renew their operations, which naturally affects their ability to produce new innovations.

The professionalization of R&D activities has provided a basis, first of all, for measuring innovation in general and, second, led us to the origins of industrial innovations. This information has been valuable for understanding where the Finnish innovations originate in our studies. As the industry-level investigation demonstrates, the nature of innovation has fulfilled a crucial role in the renewal of mature industry. Technological innovations can take the form of new products or advanced processes. The renewal of an industry then occurs through an interplay of the two types of innovation, which in addition can emerge in parallel. Moreover, the innovation level investigation of industry life cycle gives novel information to industry dynamics taking place through innovations. However, only a few innovations are so significant that they have an impact, for instance, on industry renewal and turn out to be breakthrough products.

In reality, a majority of innovations developed have a minor significance, and an even larger share of them are less viable. Failure in innovation activity is not a straightfordly understood concept in general, even among innovators themselves. Although innovation, as a term, is generally loaded with strong positive dimensions, like success, novelty and significance, it has another, less glamorous, side as well. Usually failure is realized and understood through challenge – innovation activities are about encountering and being prepared to challenge. Innovation incorporates equally failure and success; therefore, approaching innovation from the aspect of failure increases our knowledge with regard to success factors too. The vitality of innovation is, after all, judged in a market.

According to Schumpeterian theories, entrepreneurs are the key facilitators of innovations by introducing them to market. On the other hand, innovations provide companies with the core of their competitive advantage into which their prosperity is tied. Especially, in a certain phase of company's life cycle, for example, when it moves from being entrepreneurial to being a growth-oriented company, the importance of systemic management arises. Managing renewal, in which innovation is central ingredient, is therefore an important task in companies. Generally speaking, businesses constantly poise in the crossroads of current business activities and future-oriented operations, to which innovation strategies mostly relate. Consequently, recognizing and taking an advantage of opportunities is a central aspect of innovation and business activities, and at the same time makes the operational environment of entrepreneurs and innovators extremely complex.

Innovations are not developed in a vacuum; nor do companies operate in a vacuum. They are dependent on their external environment, for example, the regional innovation environment. In fact, in order to develop competitive innovations companies often need to rely on external sources of knowledge, which tend, to a certain extent, to be concentrated in a specific location. The knowledge externalities and spillovers make some regions more attractive innovation-wise, therefore being able to attract companies to establish their innovation activities in these locations. At any one moment, some areas are naturally more competitive in terms of innovation activity than others. According to our studies, it seems that innovations based on new and emerging technologies get support from embeddedness in vibrant locations. However, not all innovations are developed in central areas; rural/periphery areas prove to be very prosperous as well in producing innovations. On the other hand, it is likely that regions lose their competitive advantage in industrial composition of innovative activities after certain period of time. The same implies for innovation, for instance, in technological terms – innovations converge closer to each other. The competitive advantage innovation provides is more crucial, but at the same time more difficult, to maintain. This applies to countries, regions and firms as well as innovations – novelty in every respect is more difficult to attain.

Due to globalization, innovation networks, in which innovations are developed, are not any more merely local and domestic but increasingly internationally oriented. Our results show that, although companies collaborate extensively with various domestic and foreign actors during the innovation process, the majority of their collaborative partners are domestic. The inputs into local embeddedness and the developing regional innovation environment are worth taking into account, since it seems that international knowledge sourcing in innovation is still in a minor role compared to domestic interorganizational collaboration.

Although companies have succeeded in developing and commercializing an innovation, business may still not be successful. Ultimately, the role of companies is to bring innovations into the market for testing, since users and buyers are the last to judge whether innovation is successful or not. This trial period is likely to take a couple of years, after which a decision regarding the success of the innovation, and sometimes even the company, is made.

Innovations are becoming more and more complex, and new dimensions being added, as users get more demanding. One of the trends is the service aspect, which is more often coupled with technological product innovations, upon which the majority of our analyses touched. Even though service innovations have been identified since the early days of the study of innovations, that is, Schumpeter's writings, only recently have they been gaining more recognition and attention among innovation studies scholars as well.

Changes in innovation processes do not occur only on one level, but, as the chapters of this book indicate, change appears concurrently in innovation, company and innovation-environment levels. Therefore, we believe that our multilevel analysis has enriched our understanding on how innovations occur and change happens. However, at the same time, these studies have proven, once again, the complexity of innovation and innovation activities in general. Not just the various fresh perspectives used but also utilization of several research methodologies have similarly provided us a means to understand changes in innovation processes more deeply.

Challenges and possibilities of innovation studies

There are couple of central themes emerging for future studies on innovation and especially studies applying innovation-based data. First, there is a need for comparative studies on innovation in different countries. The results at hand, based on Finnish data, give novel insights to various aspects of innovation, but databases applying the object approach are needed in order to analyze the generalization of the results, as well as other aspects of innovation that are not studied here. Anyhow, the innovation data based on the object approach provides more accurate and reliable information on innovations and changes in innovation patterns compared to currently collected firm-level information in several countries. In the end, these two approaches complement each other and enhance the richness of perspectives used in innovation studies.

Second, we would like to stress the issue of what is and what should be considered as innovation. Different types of innovations have been recognized, and the definition of innovation is used to describe changes in a variety of phenomena: for instance, services, marketing, organizational activities and management activities are noted as a type of innovation, and even social innovations are labelled. The widening of the definition imposes on us, as innovation researchers, new challenges as old research methodologies are not necessarily accurate any more to analyze these new variations. However, new opportunities will also arise, which provide new and interesting phenomena to be investigated. The broadened view on innovation can be seen in studies on innovation, which are manifold; in February 2009 Google Scholar found 3.36 million hits for articles with 'innovation' in their title. Eventually, all this progress will result in a more thorough understanding of the innovation phenomenon.

In general, innovation is dependent on, most of all, the state of art at the current time, but it is also context-dependent. The main dimension of innovation, that is, novelty, is seen to differ between individuals and business actors, as well as by cultures and societies. Innovation in the 1920s is far from novelty in the 21st century, and innovation in developing countries might not be accounted as bringing newness in Finland, for example. Thus, an important question now and in the future in studying the phenomenon of innovation is: what do we mean by it? Should the definition change along with the changing environment we live in, or is it meaningful to preserve a more specified definition to maintain a unified understanding of the issue? However, these questions are to be answered in the future studies carried out in this intriguing research field.

References

Cantwell, John (1999) 'Innovation as the Principal Source of Growth in the Global Economy', in Daniele Archibugi, Jeremy Howells and Jonathan Michie (eds), *Innovation Policy in a Global Economy*, Cambridge: Cambridge University Press, pp. 225–41.

Drejer, I. (2004) 'Identifying Innovation in Surveys of Services: A Schumpeterian Perspective', *Research Policy*, Vol. 33, No. 3, pp. 551–62.

Fagerberg, Jan (2006) 'Innovation: A Guide to the Literature', in J. Fagerberg, D. Mowery and R. Nelson (eds), *The Oxford Handbook of Innovation*, Oxford: Oxford University Press, pp. 1–26.

Freeman, C. (1987) *Technology Policy and Economic Performance: Lessons from Japan*, London: Pinter.

Garcia, Rosanna and Calantone, Roger (2002) 'A Critical Look at Technological Innovation Typology and Innovativeness Terminology: A Literature Review', *Journal of Product Innovation Management*, Vol.19, No. 2, pp. 110–32.

Godin, Benoît (2008) 'Innovation: The History of Category', Project on the Intellectual History of Innovation, Working Paper No.1. Available at: http://www. csiic.ca/PDF/IntellectualNo1.pdf

Henderson, R. and Clark, K. (1990) 'Architectural Innovation: The Reconfiguration of Existing Product Technologies and the Failure of Established Firms', *Administrative Science Quarterly*, Vol. 35, pp. 9–30.

Johannessen, Jon-Arild, Olsen, Bjorn and Lumpkin, G. T. (2001) 'Innovation as Newness: What is New, How New, and New to Whom?', *European Journal of Innovation Management*, Vol. 4, No. 1, pp. 20–31.

Kogut, Bruce and Zander, Udo (1992) 'Knowledge of the Firm, Combinative Capabilities, and the Replication of Technology', *Organization Science*, Vol. 3, No. 3, pp. 383—97.

Marquis, Donald G. (1988) 'The Anatomy of Successful Innovations', In M. L. Tushman and W. L. Moore (eds), *Readings in the Management of Innovation* (2nd edn), USA: Ballinger, pp. 79–87.

Nelson, R. R. and Winter, S. G. (1982) *An Evolutionary Theory of Economic Change*, Cambridge, MA: Belknap Press of Harvard University Press.

OECD (2005) *Oslo Manual: Guidelines for Collecting and Interpreting Innovation Data* (3rd edn), Paris: OECD and Eurostat.

Rogers, Everett M. (1982) *Diffusion of Innovations* (3rd edn), New York: Free Press.

Rothwell, R. (1994) 'Industrial Innovations: Success, Strategy, Trends', in M. Dodgson and R. Rothwell (eds), *The Handbook of Industrial Innovation*, Cheltenham, Brookfield: Edward Elgar.

Schumpeter, Joseph A. (1963) *The Theory of Economic Development* (3rd edn), New York: Oxford University Press.

Schmookler, Jacob (1966) *Invention and Economic Growth*, Cambridge, MA: Harvard University Press.

Tidd, J., Bessant, J. and Pavitt, K. (1997) *Managing Innovation: Integrating Technological, Market and Organizational Change*, Chichester: John Wiley.

Index